The Stem Cell Microenvironment and Its Role in Regenerative Medicine and Cancer Pathogenesis

RIVER PUBLISHERS SERIES IN RESEARCH AND BUSINESS CHRONICLES: BIOTECHNOLOGY AND MEDICINE

Volume 7

Combining a deep and focused exploration of areas of basic and applied science with their fundamental business issues, the series highlights societal benefits, technical and business hurdles, and economic potentials of emerging and new technologies. In combination, the volumes relevant to a particular focus topic cluster analyses of key aspects of each of the elements of the corresponding value chain.

Aiming primarily at providing detailed snapshots of critical issues in biotechnology and medicine that are reaching a tipping point in financial investment or industrial deployment, the scope of the series encompasses various specialty areas including pharmaceutical sciences and healthcare, industrial biotechnology, and biomaterials. Areas of primary interest comprise immunology, virology, microbiology, molecular biology, stem cells, hematopoiesis, oncology, regenerative medicine, biologics, polymer science, formulation and drug delivery, renewable chemicals, manufacturing, and biorefineries.

Each volume presents comprehensive review and opinion articles covering all fundamental aspect of the focus topic. The editors/authors of each volume are experts in their respective fields and publications are peer-reviewed.

For a list of other books in this series, visit www.riverpublishers.com

The Stem Cell Microenvironment and Its Role in Regenerative Medicine and Cancer Pathogenesis

Editors

Cristian Pablo Pennisi

Aalborg University
Denmark

Mayuri Sinha Prasad

Indiana University-Purdue University Indianapolis
USA

Pranela Rameshwar

Rutgers University
USA

Published 2017 by River Publishers
River Publishers
Alsbjergvej 10, 9260 Gistrup, Denmark
www.riverpublishers.com

Distributed exclusively by Routledge
4 Park Square, Milton Park, Abingdon, Oxon OX14 4RN
605 Third Avenue, New York, NY 10158

First published in paperback 2024

The Stem Cell Microenvironment and Its Role in Regenerative Medicine and Cancer Pathogenesis / by Cristian Pablo Pennisi, Mayuri Sinha Prasad, Pranela Rameshwar.

ISBN: 978-87-93379-93-0 (hbk)
ISBN: 978-87-7004-448-6 (pbk)
ISBN: 978-1-003-33977-9 (ebk)

DOI: 10.1201/9781003339779

Contents

Preface

How stem cells behave is very much a factor of their local microenvironment, also known as the stem cell niche. Physical, chemical, or electrical signals from the neighboring cells or biochemical signals from distant cells are crucial in the cell fate decision process. A major challenge of tissue engineering is to mimic the natural cell environment by designing very sophisticated scaffolds able not only to mechanically support cells, but also to release signals biologically relevant for governing stem cell fate. In addition, increasing evidence suggests that abnormal interaction of stem cells with their niche is responsible for altered cell function leading to malignant transformation.

This book emerges as a result of the scientific contributions presented during the fifth Disputationes Workshop held in Aalborg (Denmark) in April 2014 and discusses some of the recent advances in stem cell research that may help understanding the properties of the niche that govern stem cell fate.

Editors
Cristian Pablo Pennisi
Pranela Rameshwar
Mayuri Sinha Prasad

List of Contributors

Chyan-Jang Lee, *1. Integrative Regenerative Medicine Centre, and Department of Clinical and Experimental Medicine, Linköping University, Linköping, Sweden*

Cristian Pablo Pennisi, *Laboratory for Stem Cell Research, Department of Health Science and Technology, Aalborg University, Aalborg, Denmark*

Carmen Hendricks-Guy, *BASF Corporation, 500 White Plains Road, Tarrytown, NY 10591, USA*

Chris Bath, *1. Department of Ophthalmology, Aalborg University Hospital, 9000 Denmark*
2. Laboratory for Stem Cell Research, Aalborg University, 9220 Aalborg, Denmark

Danson Muttuvelu, *Department of Ophthalmology, Aalborg University Hospital, 9000 Denmark*

Fang Wang, *Laboratory for Stem Cell Research, Aalborg University, Fredrik Bajers Vej 3B, 9220 Aalborg, Denmark*

Francesca Pagliari, *1. Physical Science and Engineering Division, King Abdullah University of Science and Technology, Jeddah, Saudi Arabia*
2. Laboratory of Molecular and Cellular Cardiology (LCMC), Department of Clinical Sciences and Translational Medicine, University of Rome "Tor Vergata", Rome, Italy

Garima Sinha, *New Jersey Medical School, Department of Medicine – Division of Hematology/Oncology, Rutgers School of Biomedical Health Science, Newark, NJ 07103, USA*

Harikrishna Nakshatri, *Departments of Surgery, Biochemistry and Molecular Biology, Indiana University Simon Cancer Center, Indiana University School of Medicine, Indianapolis, Indiana, USA*

Henrik Vorum, *Department of Ophthalmology, Aalborg University Hospital, 9000 Denmark*

John Rasmussen, *The AnyBody Research Group, Department of Mechanical and Manufacturing Engineering, Aalborg University, Aalborg, Denmark*

Jeppe Emmersen, *Laboratory for Stem Cell Research, Aalborg University, Fredrik Bajers Vej 3B, 9220 Aalborg, Denmark*

Jesper Hjortdal, *Department of Ophthalmology, Aarhus University Hospital, 8000 Aarhus, Denmark*

Lauren S. Sherman, *New Jersey Medical School, Department of Medicine–Division of Hematology/Oncology, Rutgers School of Biomedical Health Science, Newark, NJ 07103, USA*

Liliana Craciun, *BASF Corporation, 500 White Plains Road, Tarrytown, NY 10591, USA*

Linda Pilgaard, *Laboratory for Cancer Biology, Institute of Health Science and Technology, Aalborg University, Denmark*

Luciana Carosella, *Institute of Internal Medicine and Geriatrics, Catholic University of the Sacred Heart, Rome, Italy*

May Griffith, *1. Integrative Regenerative Medicine Centre, and Department of Clinical and Experimental Medicine, Linköping University, Linköping, Sweden*

Meg Duroux, *Laboratory for Cancer Biology, Institute of Health Science and Technology, Aalborg University, Denmark*

Oleksiy Buznyk, *1. Integrative Regenerative Medicine Centre, and Department of Clinical and Experimental Medicine, Linköping University, Linköping, Sweden*
2. Department of Eye Burns, Ophthalmic Reconstructive Surgery, Keratoplasty & Keratoprosthesis, Filatov Institute of Eye Diseases and Tissue Therapy, Odessa, Ukraine

Oleta A. Sandiford, *New Jersey Medical School, Department of Medicine – Division of Hematology/Oncology, Rutgers School of Biomedical Health Science, Newark, NJ 07103, USA*

Paolo Di Nardo, *1. Laboratory of Molecular and Cellular Cardiology (LCMC), Department of Clinical Sciences and Translational Medicine, University of Rome "Tor Vergata", Rome, Italy*
2. Center for Regenerative Medicine, University of Rome Tor Vergata, Italy

Pia Olsen, *1. Laboratory for Cancer Biology, Institute of Health Science and Technology, Aalborg University, Denmark*
2. Aalborg University Hospital, Department for Neuro Surgery, Denmark

Pranela Rameshwar, *New Jersey Medical School, Department of Medicine – Division of Hematology/Oncology, Rutgers School of Biomedical Health Science, Newark, NJ 07103, USA*

Stavros Papaioannou, *Laboratory for Stem Cell Research, Department of Health Science and Technology, Aalborg University, Aalborg, Denmark*

Sarah A. Bliss, *New Jersey Medical School, Department of Medicine – Division of Hematology/Oncology, Rutgers School of Biomedical Health Science, Newark, NJ 07103, USA*

Steven J. Greco, *Department of Medicine, Division of Hematology/Oncology, Rutgers School of Health Sciences, Newark, NJ 07103, USA*

Sufang Yang, *1. Laboratory for Stem Cell Research, Aalborg University, 9220 Aalborg, Denmark*
2. Animal Reproduction Institute, Guangxi University, China

Ted Deisenroth, *BASF Corporation, 500 White Plains Road, Tarrytown, NY 10591, USA*

Trine Fink, *Laboratory for Stem Cell Research, Aalborg University, 9220 Aalborg, Denmark*

Vipul Nagula, *New Jersey Medical School, Department of Medicine – Division of Hematology/Oncology, Rutgers School of Biomedical Health Science, Newark, NJ 07103, USA*

Vladimir Zachar, *Laboratory for Stem Cell Research, Department of Health Science and Technology, Aalborg University, 9220 Aalborg, Denmark*

List of Figures

List of Tables

List of Abbreviations

CTS	Cyclic Tensile Strain
ECM	ExtraCellular Matrix
FAK	Focal Adhesion Kinases
FGF	Fibroblast Growth Factor
IGFs	Insulin-like Growth Factors
MRF	Myogenic Regulatory Factor
MSCs	Mesenchymal Stem Cells
TGFβ	Transforming Growth Factor Beta
ABCG2	ATP-binding cassette subfamily G member 2
CK3	Cytokeratin 3
CFE	Colony Forming Efficiency
CLAU	Conjunctival Limbal AUtografts
CLET	Cultured Limbal Epithelial Transplantation
KLAL	KeratoLimbal Alografts
HIFs	Hypoxia Inducible Factors
HPCLK	Homologous Central Limbal Keratoplasty
LESCs	Limbal Epithelial Stem Cells
LECs	Limbal Epithelial Cells
LSCD	Limbal Stem Cell Deficiency
P63α	Tumor protein 63 isoform alpha
PMCs	Post Mitotic Cells
TACs	Transient Accelerating Cells
TDCs	Terminally Differentiated Cells
WHO	World Health Organization

1

Selective Expansion of Limbal Epithelial Stem Cells in Culture Using Hypoxia

Chris Bath[1,2], Sufang Yang[2,3], Danson Muttuvelu[1], Trine Fink[2], Jeppe Emmersen[2], Henrik Vorum[1], Jesper Hjortdal[4] and Vladimir Zachar[2]

[1]Department of Ophthalmology, Aalborg University Hospital, 9000 Denmark
[2]Laboratory for Stem Cell Research, Aalborg University, Fredrik Bajers Vej 3B, 9220 Aalborg, Denmark
[3]Animal Reproduction Institute, Guangxi University, China
[4]Department of Ophthalmology, Aarhus University Hospital, 8000 Aarhus, Denmark

Abstract

Limbal epithelial stem cells (LESCs) maintain the corneal epithelium throughout life and are crucial for both corneal integrity and vision. In this study, LESCs were expanded in either a culture system using 3T3 feeder cells in growth medium supplemented with serum, or in a culture system without feeder cells using commercially available serum-free medium (EpiLife). Cells were maintained at an ambient oxygen concentration of 20% or at various levels of hypoxia (15%, 10%, 5%, and 2%) throughout the period of expansion. The effect of ambient oxygen concentration on growth, cell cycle, colony forming efficiency (CFE), and expression of stem cell markers ABCG2 and p63α and differentiation marker CK3 were determined at different time points. Low oxygen levels were found to maintain a stem cell phenotype with low proliferative rate, high CFE, and high expression of ABCG2 and p63α as well as low expression of CK3. The relation between degree of differentiation and ambient oxygen concentration in the culture system seems to mirror the natural environment of the limbal niche. Hypoxic culture could therefore potentially improve stem cell grafts for cultured limbal epithelial transplantation (CLET).

Keywords: Limbus cornea, Adult stem cells, Regenerative medicine, Cell hypoxia, Primary cell culture.

1.1 Introduction

Vision is important for normal quality of life and blindness is universally feared. The World Health Organization (WHO) has estimated that the prevalence of blindness is around 39 million people worldwide with up to 285 million people having impaired vision [1]. Corneal opacities accounts for 4% of blind cases and around 1% of impaired vision cases, and treatment for these causes of decreased visual acuity has traditionally relied on tissue donations for either keratoplasty or limbal tissue transplantation. Regenerative ophthalmic medicine holds great promise to deliver new treatment regimens to address current challenges such as scarcity of donor materials.

The cornea can be regarded as the window of the eye as it allows light to enter into the eye to reach the phototransducing cells of the retina. This function is crucially dependent on intact transparency of corneal tissue, which can be disturbed in a variety of clinical settings. The cornea and associated tear film are furthermore responsible for almost two-thirds of the total refractive power of the eye [2] and hold important protective properties for inner ocular structures.

The human cornea is composed of three cellular layers and two interface layers. The superficial cellular layer facing the external environment is a multilayered, non-keratinized squamous epithelium [3], that is separated from the corneal stroma by a thin acellular layer termed Bowman's membrane. The corneal stroma accounts for 90% of the total corneal thickness of 515 ± 33 μm [4], and is composed of regularly arranged collagen lamellae that promotes transparency by removal of light scatter in a process of destructive interference [5]. The innermost cellular layer of the cornea is the corneal endothelium, that is composed of a monolayer of non-regenerative cells *in vivo* that maintains corneal clarity by both a barrier function, ascribed to intercellular tight junctions, but also an active pump function using Na^+/K^+ ATPase to regulate hydration levels of the corneal stroma [6]. Corneal transparency and thereby function is dependent on the integrity of all corneal cellular layers but also health of ocular adnexa. Diseases involving one or more cellular layers of the cornea give rise to decreased visual acuity and debilitating symptoms in patients.

In recent years, the field of regenerative ophthalmic medicine has experienced great progress towards curing corneal diseases affecting one or more of the cellular layers using bioengineered tissue replacements. As mentioned below, Pellegrini and co-authors pioneered the treatment of corneal epithelial diseases by transplantation of *ex vivo* expanded epithelial stem cells [7]. Bioengineering techniques are rapidly developing that enable surgeons to perform keratoplastic procedures using artificially crafted stromal replacements [8]. Finally, corneal endothelial substitutes can treat diseases of the innermost endothelial cell layer that impede the pump function resulting in corneal edema [6].

1.2 Clinical Application of Bioengineered Corneal Epithelial Stem Cell Grafts

The corneal epithelium is a multi-layered, non-keratinized squamous epithelium that is continuously regenerated by dedicated unipotent tissue-specific stem cells. These stem cells are termed limbal epithelial stem cells (LESCs), and are located in specific stem cell niches in the corneal limbus called limbal crypts [9], limbal epithelial crypts [10] and focal stromal projections [10]. From this nurturing micro-environment, LESCs divide by symmetrical or asymmetrical cell division to give rise to a population of more committed progenitors termed transient accelerating cells (TACs) that migrate centrally and superficially to differentiate into postmitotic cells (PMCs) and ultimately to terminally differentiated cells (TDCs) that are continuously lost to the external environment (Figure 1.1). This spatially unique differentiation scheme for corneal epithelial homeostasis has been mathematically explained in the so-called X, Y, Z hypothesis of corneal epithelial regeneration [11]. Various diseases, most often grouped into acquired, hereditary, iatrogenic and idiopathic causes [12], give rise to defect or dysfunctional LESCs thereby disrupting epithelial homeostasis resulting in decreased visual acuity and pain in patients. This disease of the corneal epithelial stem cell population has been termed limbal stem cell deficiency (LSCD) and treatment requires transplantation of LESCs.

Traditional keratoplasty cannot cure these patients, as only the central button being devoid of LESCs is transplanted to the recipient. Historically, LESCs have been transplanted using whole tissue blocks from donors by homologous penetrating central limbal keratoplasty (HPCLK) [13], keratolimbal

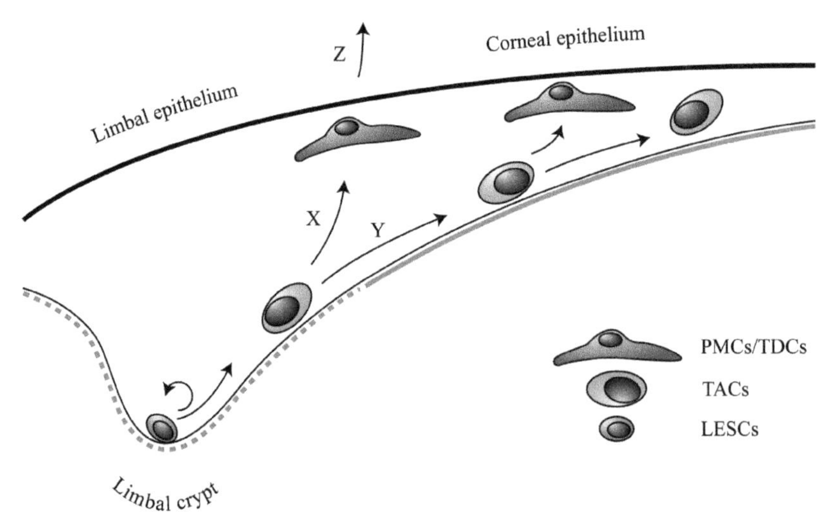

Figure 1.1 The X, Y, Z hypothesis of corneal epithelial maintenance. Stem cells (LESCs) are capable of symmetrical/asymmetrical cell division and differentiate into TACs during centripetal cell migration (Y). Transient accelerating cells mature into PMCs/TDCs during superficial migration (X) and are ultimately sloughed off to external environment (Z). Epithelial homoeostasis requires X + Y = Z. Green line represents basement membrane. LESCs, limbal epithelial stem cells; TACs, transient accelerating cells; PMCs, postmitotic cells; TDCs, terminal differentiated cells. Reproduced from [3] with permission from Wiley Blackwell.

allografts (KLAL) or conjunctival limbal autografts (CLAU) [14]. Recently, researchers have focused on curing LSCD by transplantation of bioengineered tissues containing LESCs to patients in a process termed cultured limbal epithelial transplantation (CLET) [14]. CLET was originally performed by Pellegrini and co-authors [7] using culture techniques analogous to epidermal research [15, 16]. The process of CLET is outlined in Figure 1.2. Briefly, the technique includes acquiring a small biopsy containing LESCs from either a donor eye in cases of bilateral LSCD or the contralateral eye of the recipient in cases of unilateral disease. These cells are subsequently expanded in the laboratory on a dedicated carrier like e.g. amniotic membrane. The bioengineered transplant containing LESCs is then transferred to the surgical facility, where the recipient eye is prepared by debridement of pannus. The bioengineered stem cell transplant is finally placed on the corneal wound bed to enable recreation of a stable corneal epithelium. In a subsequent surgical procedure it is possible to replace opaque stroma in more profound disease by either conventional keratoplasty or by using artificial stromal replacements [8]. CLET has theoretical advantages compared to traditional

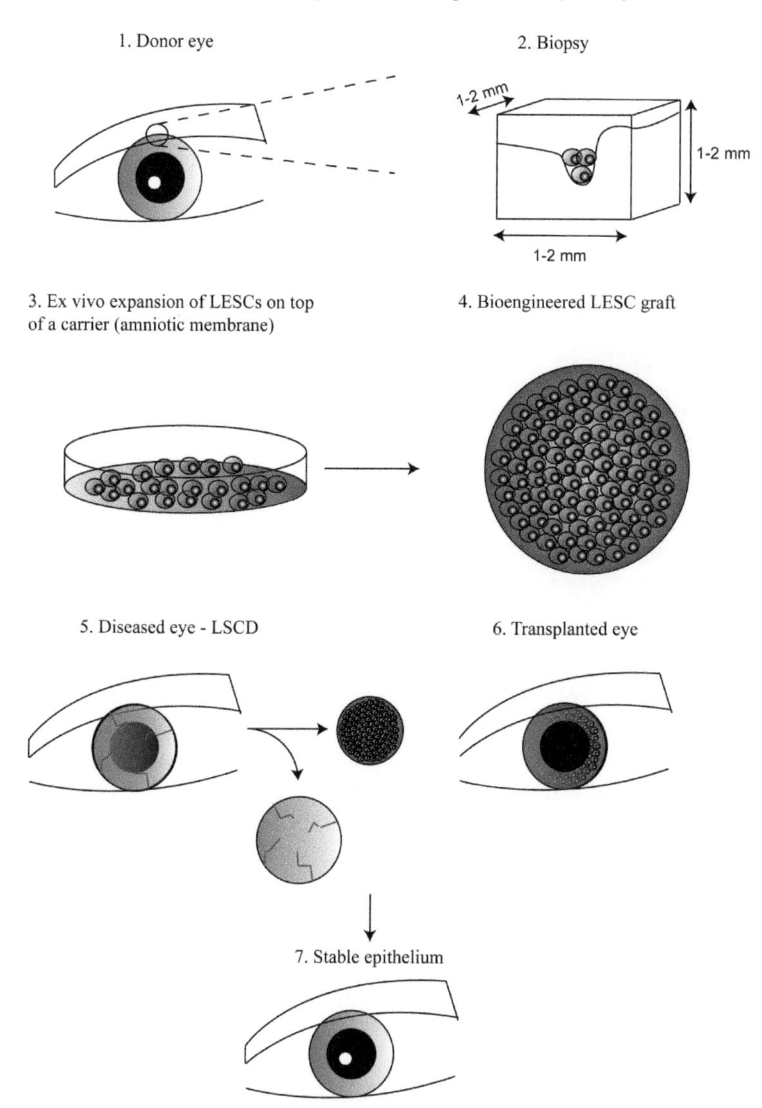

Figure 1.2 Procedure of cultivated limbal epithelial transplantation (CLET). (1) A biopsy containing LESCs is obtained from healthy corneal limbus. (2) LESCs in limbal crypts are enzymatically isolated from a small biopsy of 1–2 mm for *ex vivo* expansion. (3–4) A monolayer of limbal epithelial cells expanded *ex vivo* on a carrier constitutes a bioengineered stem cell graft. (5) Recipient eye is prepared by debridement of diseased pannus tissue. (6–7) Placing the graft on the wound bed recreates a stable surface epithelium. CLET, cultivated limbal epithelial transplantation; LESCs, limbal epithelial stem cells; LSCD, limbal stem cell deficiency.

tissue transfer techniques like KLAL or CLAU such as reduced risk of iatrogenic LSCD in the donor eye, diminished risk of rejection due to lack of Langerhans´cells within the graft and the possibility of repeating surgical procedure in case of initial failure [3]. Optimisation of culture techniques though seem important, as it has been shown that the likelihood of success after CLET is substantially increased if the cultured graft contains above 3% ΔNP63α positive cells [17], which is a widely accepted stem cell biomarker for LESCs [18].

1.3 Culture Techniques for Bioengineered Stem Cell Sheets

Currently, two major culture techniques exist for propagation of LESCs *ex vivo* prior to CLET. The explant system expands cells by placing small tissue blocks measuring 1–2 mm containing LESC niche structures on a suitable substrate allowing subsequent outgrowth of cells during the culture process. The so-called "dissociation" system enzymatically dissociates cells from a tissue sample prior to seeding cells on a suitable substrate for *ex vivo* expansion. A direct comparison of these methods for crafting bioengineered cell sheets for CLET seem to support that the dissociation system is superior to the explant system in expanding LESCs [19]. Some authors have suggested that the explant system mainly allows outgrowth of TACs, and it has also been suggested that progenitors could be lost in a process of epithelial-mesenchymal transition [20]. Optimal enzymatic dissociation of cells in the dissociation system seem to be performed using a combination of dispase II and trypsin/EDTA [21], and evidence using scanning electron microscopy supports that also LESCs located in deep niche structures are harvested by this enzymatic treatment [19]. Since the advent of CLET, much emphasis has been placed on the optimisation of various culture variables to improve outcome of surgery, but only recently has emphasis been placed on the gaseous environment of cultures [22].

1.4 Selective Expansion of LESCs Using Hypoxia

Increasing awareness of oxygen as a regulator of stem cell growth and differentiation has prompted investigators to rename traditional hypoxic culture conditions as *in situ* normoxia [23] or physiological normoxia [24], as these gaseous conditions reflect the expected milieu encountered by stem cells in their niches. Hypoxia has been shown to maintain various stem cells in culture [25, 26], most likely attributable to a concerted action by hypoxia inducible factors (HIFs) (reviewed in [27]). Kwan and co-authors performed measurements of oxygen concentrations in various layers of the central cornea in rabbits breathing atmospheric oxygen, and were able to show that oxygen decreased from 123$\pm$10 mmHg below the tearfilm to 65 mmHg at the epitheliostromal junction [28]. Several factors could easily contribute to an even lower oxygen concentration in stem cell niches at the corneal limbus such as human corneal epithelium being thicker than rabbit epithelium [29] and the limbal epithelium having more cellular layers than central corneal epithelium [30]. Furthermore, a nocturnal decrease in oxygen availability during sleep due to eye lid closure could easily increase hypoxia in the limbal epithelium [31], and as oxygen concentrations in most peripheral capillaries are 0.5–2.6%, the perilimbal vascular arcades are not expected to change the hypoxic nature of the niche.

Using a dedicated hypoxic cell culture facility (BioSpherix Xvivo facility, http://www.biospherix.com/equipment/cytocentric/systems.html), we have recently tested the effect of different oxygen concentrations of 2%, 5%, 10%, 15%, and 20% on the growth and differentiation status of limbal epithelial cells (LECs) in the dissociation culture system using two commonly used media formulations [22]. Cultures were analyzed using advanced computerized fluorescence microscopy and software analysis on whole populations of LECs after expansion in either a culture system employing γ-irradiated 3T3 cells as feeder cells and serum containing medium or in a newer system devoid of feeder cells using a commercially available serum-free semi-defined medium called EpiLife. Expanded cells exhibited a LESC phenotype in hypoxic culture conditions of 2–5% with slow growth, high colony forming efficiency, high expression of presumed stem cell markers tumor protein 63 isoform α (p63α) and ATP-binding cassette subfamily G member 2 (ABCG2) and low expression of differentiation marker cytokeratin 3 (CK3). Conversely, cells expanded in 15% O_2 exhibited slow growth, low colony forming efficiency, low expression of p63α and ABCG2, and high expression of CK3. Interestingly, 15% O_2 resembles the oxygen concentration measured by Kwan and co-authors in the differentiation compartment directly below the tear film [28]. An intermediate O_2 concentration of 10% revealed cellular phenotypes resembling transient accelerating cells with fast growth, intermediate expression levels of both stem cell and differentiation markers as well as intermediate level of colony forming efficiency. The upregulation of particular genes related to hypoxia in the deep limbal crypts has been shown using laser capture microdissection and RNA-sequencing [32].

1.5 Future Perspectives

Being able to reproduce cellular phenotypes in culture, as they exist along their spatially defined differentiation pathway *in vivo*, by controlled atmospheric oxygen levels could have great implications for the optimization of bioengineered stem cell grafts used in the CLET procedure, as it has been shown that the success ratio of surgery is critically dependent on a high fraction of stem cells contained within the graft [17]. Novel regenerative therapies within ophthalmology using tissue replacements hold promise to circumvent current shortcomings of therapy caused by scarcity of donor materials.

References

[1] Pascolini, D., Mariotti, S.P., 2011. Global estimates of visual impairment: 2010. Br J Ophthalmol.

[2] Ruberti, J.W., Roy, A.S., Roberts, C.J., 2011. Corneal biomechanics and biomaterials. Annu Rev Biomed Eng 13, 269–295.

[3] Bath, C., 2013. Human corneal epithelial subpopulations: oxygen dependent ex vivo expansion and transcriptional profiling. Acta Ophthalmol 91 Thesis 4, 1–34.

[4] Olsen, T., Ehlers, N., 1984. The thickness of the human cornea as determined by a specular method. Acta Ophthalmol 62, 859–871.

[5] Maurice, D.M., 1957. The structure and transparency of the cornea. The Journal of physiology 136, 263.

[6] Koizumi, N., Okumura, N., Kinoshita, S., 2012. Development of new therapeutic modalities for corneal endothelial disease focused on the proliferation of corneal endothelial cells using animal models. Exp Eye Res 95, 60–67.

[7] Pellegrini, G., Traverso, C.E., Franzi, A.T., Zingirian, M., Cancedda, R., de Luca, M., 1997. Long-term restoration of damaged corneal surfaces with autologous cultivated corneal epithelium. The Lancet 349, 990–993.

[8] Fagerholm, P., Lagali, N.S., Merrett, K., Jackson, W.B., Munger, R., Liu, Y., Polarek, J.W., Soderqvist, M., Griffith, M., 2010. A biosynthetic alternative to human donor tissue for inducing corneal regeneration: 24-month follow-up of a phase 1 clinical study. Science Translational Medicine 2, 46ra61–46ra61.

[9] Dua, H.S., 2005. Limbal epithelial crypts: a novel anatomical structure and a putative limbal stem cell niche. British Journal of Ophthalmology 89, 529–532.

[10] Shortt, A.J., Secker, G.A., Munro, P.M., Khaw, P.T., Tuft, S.J., Daniels, J.T., 2007. Characterization of the Limbal Epithelial Stem Cell Niche: Novel Imaging Techniques Permit In Vivo Observation and Targeted Biopsy of Limbal Epithelial Stem Cells. Stem Cells 25, 1402–1409.

[11] Thoft, R.A., Friend, J., 1983. The X, Y, Z hypothesis of corneal epithelial maintenance. Invest Ophthalmol Vis Sci 24, 1442–1443.

[12] Osei-Bempong, C., Figueiredo, F.C., Lako, M., 2013. The limbal epithelium of the eye–a review of limbal stem cell biology, disease and treatment. Bioessays 35, 211–219.

[13] Sundmacher, R., Reinhard, T., 1996. Central corneolimbal transplantation under systemic ciclosporin A cover for severe limbal stem cell insufficiency. Graefes Arch. Clin. Exp. Ophthalmol. 234 (Suppl 1), S122–5.

[14] Shortt, A.J., Tuft, S.J., Daniels, J.T., 2011. Corneal stem cells in the eye clinic. British Medical Bulletin 100, 209–225.

[15] Rheinwald, J.G., Green, H., 1975a. Formation of a keratinizing epithelium in culture by a cloned cell line derived from a teratoma. Cell 6, 317–330.

[16] Rheinwald, J.G., Green, H., 1975b. Serial cultivation of strains of human epidermal keratinocytes: the formation of keratinizing colonies from single cells. Cell 6, 331–343.

[17] Rama, P., Matuska, S., Paganoni, G., Spinelli, A., de Luca, M., Pellegrini, G., 2010. Limbal Stem-Cell Therapy and Long-Term Corneal Regeneration. N Engl J Med 363, 147–155.

[18] Di Iorio, E., 2005. Isoforms of Np63 and the migration of ocular limbal cells in human corneal regeneration. Proceedings of the National Academy of Sciences 102, 9523–9528.

[19] Zhang, X., Sun, H., Tang, X., Ji, J., Li, X., Sun, J., Ma, Z., Yuan, J., Han, Z., 2005. Comparison of cell-suspension and explant culture of rabbit limbal epithelial cells. Exp Eye Res 80, 227–233.

[20] Li, W., Hayashida, Y., He, H., Kuo, C.L., Tseng, S.C.G., 2007. The Fate of Limbal Epithelial Progenitor Cells during Explant Culture on Intact Amniotic Membrane. Invest Ophthalmol Vis Sci 48, 605–613.
[21] Meyer-Blazejewska, E.A., Kruse, F.E., Bitterer, K., Meyer, C., Hofmann-Rummelt, C., Wunsch, P.H., Schlötzer-Schrehardt, U., 2010. Preservation of the Limbal Stem Cell Phenotype by Appropriate Culture Techniques. Invest Ophthalmol Vis Sci 51, 765–774.
[22] Bath, C., Sufang, Y., Muttuvelu, D., Fink, T., Emmersen, J., Henrik, V., Hjortdal, J., Zachar, V., 2013b. Hypoxia is a key regulator of limbal epithelial stem cell growth and differentiation. Stem Cell Research 10, 349–360.
[23] Ivanovic, Z., 2009. Hypoxia or in situ normoxia: The stem cell paradigm. J Cell Physiol 219, 271–275.
[24] Simon, M.C., Keith, B., 2008. The role of oxygen availability in embryonic development and stem cell function. Nat. Rev. Mol. Cell Biol. 9, 285–296.
[25] Prasad, S.M., Czepiel, M., Cetinkaya, C., Smigielska, K., Weli, S.C., Lysdahl, H., Gabrielsen, A., Petersen, K., Ehlers, N., Fink, T., Minger, S.L., Zachar, V., 2009. Continuous hypoxic culturing maintains activation of Notch and allows long-term propagation of human embryonic stem cells without spontaneous differentiation. Cell Prolif 42, 63–74.
[26] Sufang, Y., Pilgaard, L., Chase, L.G., Boucher, S., Vemuri, M.C., Fink, T., Zachar, V., 2012. Defined xenogeneic-free and hypoxic environment provides superior conditions for long-term expansion of human adipose-derived stem cells. Tissue engineering Part C, Methods 18, 593–602.
[27] Benizri, E., Ginouvès, A., Berra, E., 2008. The magic of the hypoxia-signaling cascade. Cell Mol Life Sci 65, 1133–1149.
[28] Kwan, M., Niinikoski, J., Hunt, T.K., 1972. In vivo measurements of oxygen tension in the cornea, aqueous humor, and anterior lens of the open eye. Invest Ophthalmol 11, 108–114.
[29] Reiser, B.J., Ignacio, T.S., Wang, Y., Taban, M., Graff, J.M., Sweet, P., Chen, Z., Chuck, R.S., 2005. In vitro measurement of rabbit corneal epithelial thickness using ultrahigh resolution optical coherence tomography. Vet Ophthalmol 8, 85–88.
[30] Feng, Y., Simpson, T.L., 2008. Corneal, limbal, and conjunctival epithelial thickness from optical coherence tomography. Optom Vis Sci 85, E880–3.

[31] Shimmura, S., Shimoyama, M., Hojo, M., Urayama, K., Tsubota, K., 1998. Reoxygenation injury in a cultured corneal epithelial cell line protected by the uptake of lactoferrin. Invest Ophthalmol Vis Sci 39, 1346–1351.

[32] Bath, C., Muttuvelu, D., Emmersen, J., Henrik, V., Hjortdal, J., Zachar, V., 2013a. Transcriptional Dissection of Human Limbal NicheCompartments by Massive Parallel Sequencing. PLoS ONE 8, e64244.

2

Mesenchymal Stem Cells and Pathotropism: Regenerative Potential and Safety Concerns

Garima Sinha, Sarah A. Bliss, Lauren S. Sherman, Oleta A. Sandiford, Vipul Nagula and Pranela Rameshwar

New Jersey Medical School, Department of Medicine – Division of Hematology/Oncology, Rutgers School of Biomedical Health Science, Newark, NJ 07103, USA

Abstract

Mesenchymal Stem cells (MSCs) are ubiquitously expressed in several organs, but the major sites in adults are bone marrow and adipose tissue. MSCs can form several cells belonging to all germ layers, such as neurons and cardiomyocytes. MSCs have the potential to be used in cell therapy for many clinical problems, e.g., tissue regeneration, replacement, and to suppress inflammatory processes. MSCs are attractive due to reduced ethical concerns, ease in expansion and ability to be used as 'off-the-shelf' cells. MSCs can be indicated for clinical disorders due to their homing to regions of high cytokines such as tissue insult. This process is generally referred as pathotropism. MSCs have been placed in numerous clinical trials. Thus far, there is no evidence of safety concerns. Besides transplantation of hematopoietic stem cells, treatments with other stem cells are relatively recent. Thus, MSC therapy requires strict monitoring for safety issues. The pathotropic effect of MSCs allows these cells to home to tumors. This property led to the use of MSCs as cellular vehicle for drugs. A major concern of using MSCs in regenerative medicine is their ability to protect and support tumor growth. This chapter focuses on the potential safety concerns of using MSCs. This issue is particularly important if the recipient of stem cells has an undiagnosed tumor or is in cancer remission.

Keywords: Stem cells, Cancer, Regeneration, Dormancy, Bone marrow, Pathotropism.

2.1 Introduction

The presence of stem cells in all organs strongly suggests that these cells may be important to protect, and perhaps replace, tissue during minimal insults. These seeming baseline properties have led scientists to investigate how stem cells can be used in tissue repair. This led to the removal of stem cells from the natural microenvironment for *ex vivo* manipulation in the absence of the natural microenvironment. There are reports in which autologous stem cells are removed and immediately transplanted (www.clinicaltrials.gov). However, in the majority of cases, the stem cells are expanded *in vitro.* This places the stem cells at risk for mutations and functional changes that could differ from their endogenous properties. As stem cells move to patients, there must be consideration on their safety and the functional alterations.

Stem cells can self-renew and differentiate into any cell type. These two major characteristics provide them with the potential to regenerate tissues and for use in organogenesis. Embryonic and induced pluripotent stem cells have scientific challenges mostly due to the ease in forming tumors. Stem cells in adults, fetus, cord, cord blood and placenta show potential in clinical application. Stem cells in the adult brain and bone marrow, such as mesenchymal stem cells (MSCs) have shown promise in regenerative medicine and in drug delivery to tumors [1].

2.2 Mesenchymal Stem Cells (MSC)

MSCs are primordial in origin and can be isolated from fetal and adult tissues such as the placenta, bone marrow and adipose tissues [2–5]. MSCs are well-characterized with regards to phenotype. There are commercially available antibodies to phenotype MSCs with antibodies such as those targeted to CD73, CD90 and CD105 [6]. MSCs do not express hematopoietic markers such as CD45. MSCs have been reported to express vimentin and fibronectin [7]. In addition to phenotype, MSCs are characterized by functionality, including assays to ensure multipotency and immune properties [6].

MSCs have significantly reduced ethical concerns, are easy to expand *in vitro*, and more importantly, can be used as 'off the shelf' sources in cell therapy. An important consideration with MSCs as cell therapy is the changing microenvironment. Thus, a clear understanding of how the changing niche affects the functions of MSCs is key. The immune function of MSCs is particularly relevant. MSCs can be immune enhancer and suppressor cells. The immune function of MSCs depends on the milieu of the microenvironment. Specific cytokines and chemokines can be chemoattractant to facilitate the migration and homing of MSCs and other immune cells to the site of tissue injury.

2.2.1 MSC Immunology

The immune functions of MSCs are interesting since these stem cells can be stimulatory as well as suppressive, depending on the microenvironment (Figure 2.1) [8]. MSCs, like other immune cells within a niche, can produce cytokines, thereby establishing communication with the cells found within

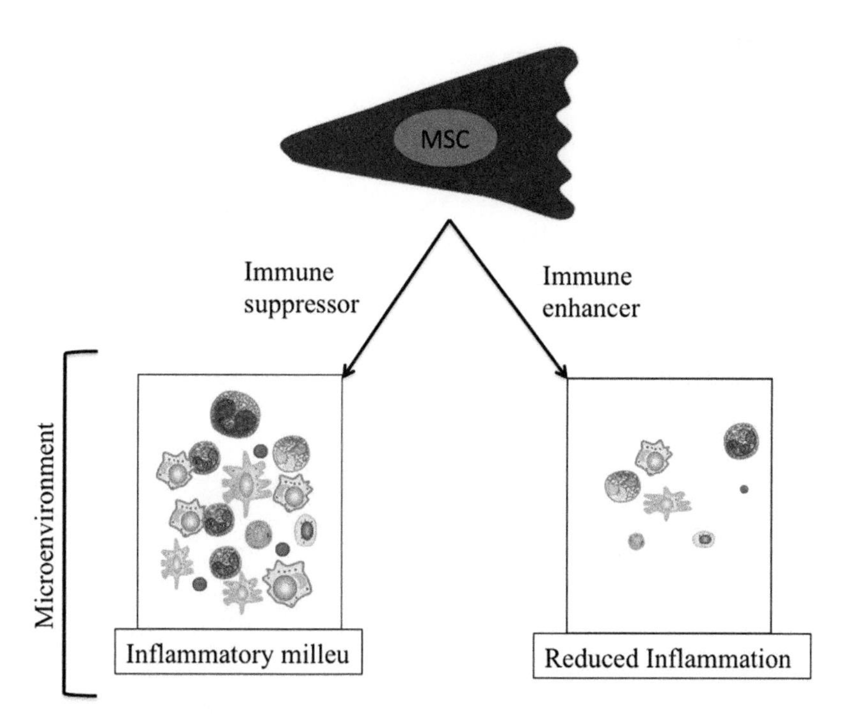

Figure 2.1 Microenvironment dictates dual role of MSCs. The inflamed microenvironment licences MSCs to immune suppressor cells whereas the reduced inflammation signal allows MSCs to be an immune enhancer.

the tissue microenvironment [9]. The expression of major histocompatibility complex-II (MHC-II) was proposed as a minimum requirement for a cell to be designated MSC [6]. There are several reports showing MSCs as negative for MHC-II, despite the correct phenotype and evidence of multipotency [10]. There are ongoing investigations to determine if the difference in MHC-II expression is due to the source from which the MSCs was derived, adipose tissue versus bone marrow, or if the difference could be explained by the type of culture condition.

MHC-II expression provides the MSCs with the ability to function as antigen presenting cells (APCs) to stimulate the immune system [11]. The method by which MSCs function as APCs is different from professional APCs. MSCs respond differently to interferon γ (IFNγ) with regards to MHC-II expression: at high levels, MHC-II expression is decreased whereas at low level, MHC-II is increased [12]. This difference is relevant for the translation of the science on MSCs because these stem cells would alter their functions with the change in the inflammatory microenvironment. Our data showed a similar re-expression of MHC-II on neurons in the presence of IFNγ [13]. This reexpression of MHC has the potential safety issues with MSC treatment.

MSCs can be licensed to become immune suppressor cells within an inflammatory milieu [14]. As immune suppressor cells, the MSCs have undetectable surface MHC-II, inhibit T-cell and B-cell stimulatory responses, and inhibit natural killer activity [6, 10, 15, 16]. These suppressive effects are generally associated with an increase in regulatory T-cells [17]. The immune properties of MSCs have been suggested to have a role in tissue repair [18].

2.3 MSC in Tumor Support

Fetal and adult MSCs can support and protect tumors [19]. It is unlikely that transplanted MSCs will form tumors since they are not likely to survive for long periods. However, they could 'waken' sleeping cancer cells since they can suppress the immune response and at the same time support tumor growth [20]. MSCs have been shown to initiate the growth and metastasis of tumor cells [21–25]. This tumor-promoting role of MSCs can occur with solid and hematologic malignancies [26–28].

The mechanisms by which MSCs support tumor growth would require further studies with different types of tumors. Also, since tumors are heterogeneous, the studies must be performed with various subsets of cancer cells.

2.3.1 MSC in Drug Delivery to Tumors

Based on the above discussion, pathotropism of MSCs could be deleterious for the recipient who might have an undiagnosed or is in remission from cancer (Figure 2.2). This disadvantage can be an advantage because the ability of MSCs to be attracted to the sites of tumor growth could be applied to have these stem cells deliver therapies to tumors. As an example, glioma cells secrete soluble factors, which act as chemoattractants for MSCs to the site of the tumor. MSCs with ectopic expression of anti-tumor molecules such as TNFα and IFN-β were studied as a method to inhibit tumor promotion [29–33].

The use of MSCs as a cellular vehicle for drugs to target glioblastoma multiforme (GBM) is an attractive area of research. GBM is an aggressive and invasive cancer with poor prognosis, despite aggressive treatments that include tumor resection with radio- and chemotherapy [34]. The utility of drug delivery has been demonstrated in rats in which the MSCs migrated to the region of gliomas [35]. In these studies, the MSCs were intracranially implanted into rats with GBM. The MSCs then migrated to and dispersed within the tumor mass [35]. A similar study with immunocompromised mice showed human MSCs migrating to the region of human gliomas [36]. In these studies, MSCs were injected into the ipsilateral and contralateral carotid arteries of the mice [36]. Other studies injected the MSCs intratumorally [37]. MSCs can be used to deliver RNA through gap junction or through exosomes, as demonstrated with the transfer of exosomes-derived anti-miR9 [38].

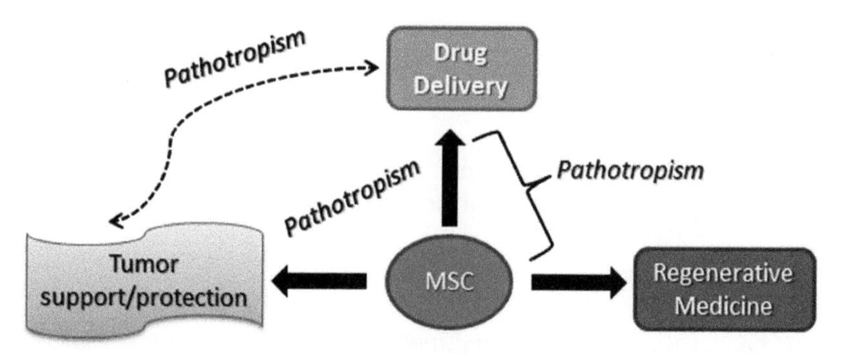

Figure 2.2 Pathotropic effects of MSCs in regenerative medicine. Shown is a MSC in the center that could be attracted to the site of tissue injury where it can be attracted to regenerate the tissue. Similarly, tumors, which can produce chemoattractants, could attract MSCs. The same pathotropic effect could be used to deliver drugs to areas of tumors.

2.4 Conclusion

A major concern with using MSCs, including the use as cellular deliverers of drugs, is the potential of MSCs to support tumor growth and to immunosuppress cells. The latter function will provide a survival advantage for the tumor while providing the MSCs with the ability to support tumor growth. At this time, there is no longitudinal study to determine if MSC treatment is safe. These are necessary studies that should be addressed. To understand how MSCs could be safely employed, it is necessary to develop experimental models of dormancy since this will recapitulate undiagnosed tumors and cancer in remission. Also, studies are needed to develop hierarchy of tumors.

A mistake noted in the literature is for scientists to presume that the properties of cancer stem cells are similar to normal stem cells. Most notable among these differences is the designation of cancer stem cells as small population SP) cells, which would indicate cells with low metabolic activity. Unlike normal stem cells, cancer stem cells are malignant cells, which would indicate relatively large cells as compared to normal stem cells. The regenerative potential of MSCs needs to be studied in conjunction with robust studies of cancer stem cells.

Regarding the drug delivery capacity of MSCs, it might be necessary to eliminate these cells after the drug is released. At present there are methods to induce suicidal genes to eliminate the MSCs [39].

References

[1] Gjorgieva, D., N. Zaidman, and D. Bosnakovski, 'Mesenchymal stem cells for anti-cancer drug delivery', Recent Pat Anticancer Drug Discov, 2013. **8**(3): 310–8.

[2] Campagnoli, C., et al., 'Identification of mesenchymal stem/progenitor cells in human first-trimester fetal blood, liver, and bone marrow', Blood, 2001. **98**(8): 2396–402.

[3] He, Q., C. Wan, and G. Li, 'Concise review: multipotent mesenchymal stromal cells in blood', Stem Cells, 2007. **25**(1): 69–77.

[4] Lee, O.K., et al., 'Isolation of multipotent mesenchymal stem cells from umbilical cord blood', Blood, 2004. **103**(5): 1669–75.

[5] Tsuda, H., et al., 'Allogenic fetal membrane-derived mesenchymal stem cells contribute to renal repair in experimental glomerulonephritis', Am J Physiol Renal Physiol, 2010. **299**(5): F1004–13.

[6] Dominici, M., et al., 'Minimal criteria for defining multipotent mesenchymal stromal cells', The International Society for Cellular Therapy position statement. Cytotherapy, 2006. **8**(4): 315–7.

[7] Vogel, W., et al., 'Heterogeneity among human bone marrow-derived mesenchymal stem cells and neural progenitor cells', Haematologica, 2003. **88**(2): 126–33.

[8] Sherman, L.S., et al., 'Moving from the laboratory bench to patients' bedside: considerations for effective therapy with stem cells', Clin Transl Sci, 2011. **4**(5): 380–6.

[9] Castillo, M., et al., 'The immune properties of mesenchymal stem cells', Int J Biomed Sci, 2007. **3**(2): 76–80.

[10] Jacobs, S.A., et al., 'Immunological characteristics of human mesenchymal stem cells and multipotent adult progenitor cells', Immunol Cell Biol, 2013. **91**(1): 32–9.

[11] Romieu-Mourez, R., et al., 'Regulation of MHC class II expression and antigen processing in murine and human mesenchymal stromal cells by IFN-gamma, TGF-beta, and cell density', J Immunol, 2007. **179**(3): 1549–58.

[12] Tang, K.C., et al., 'Down-regulation of MHC II in mesenchymal stem cells at high IFN-gamma can be partly explained by cytoplasmic retention of CIITA', J Immunol, 2008. **180**(3): 1826–33.

[13] Cheng, Z., et al., 'Targeted migration of mesenchymal stem cells modified with CXCR4 gene to infarcted myocardium improves cardiac performance', Mol Ther, 2008. **16**(3): 571–9.

[14] English, K. and B.P. Mahon, 'Allogeneic mesenchymal stem cells: agents of immune modulation', J Cell Biochem, 2011. **112**(8): 1963–8.

[15] Corcione, A., et al., 'Human mesenchymal stem cells modulate B-cell functions', Blood, 2006. **107**(1): 367–72.

[16] De Miguel, M.P., et al., 'Immunosuppressive properties of mesenchymal stem cells: advances and applications', Curr Mol Med, 2012. **12**(5): 574–91.

[17] Maccario, R., et al., 'Interaction of human mesenchymal stem cells with cells involved in alloantigen-specific immune response favors the differentiation of CD4+ T-cell subsets expressing a regulatory/suppressive phenotype', Haematologica, 2005. **90**(4): 516–25.

[18] Hoogduijn, M.J., et al., 'The immunomodulatory properties of mesenchymal stem cells and their use for immunotherapy', Int Immunopharmacol, 2010. **10**(12): 1496–500.

[19] Zhu, W., et al., 'Mesenchymal stem cells derived from bone marrow favor tumor cell growth *in vivo*', Exp Mol Pathol, 2006. **80**(3): 267–74.
[20] Zhang, T., et al., 'Bone marrow-derived mesenchymal stem cells promote growth and angiogenesis of breast and prostate tumors', Stem Cell Res Ther, 2013. **4**(3): 70.
[21] Djouad, F., et al., 'Immunosuppressive effect of mesenchymal stem cells favors tumor growth in allogeneic animals', Blood, 2003. **102**(10): 3837–44.
[22] Goldstein, R.H., et al., 'Human bone marrow-derived MSCs can home to orthotopic breast cancer tumors and promote bone metastasis', Cancer Res, 2010. **70**(24): 10044–50.
[23] Karnoub, A.E., et al., 'Mesenchymal stem cells within tumour stroma promote breast cancer metastasis', Nature, 2007. **449**(7162): 557–63.
[24] Shinagawa, K., et al., 'Mesenchymal stem cells enhance growth and metastasis of colon cancer', Int J Cancer, 2010. **127**(10): 2323–33.
[25] Yu, J.M., et al., 'Mesenchymal stem cells derived from human adipose tissues favor tumor cell growth *in vivo*', Stem Cells Dev, 2008. **17**(3): 463–73.
[26] Ame-Thomas, P., et al., 'Human mesenchymal stem cells isolated from bone marrow and lymphoid organs support tumor B-cell growth: role of stromal cells in follicular lymphoma pathogenesis', Blood, 2007. **109**(2): 693–702.
[27] Dubois, S.G., et al., 'Isolation of human adipose-derived stem cells from biopsies and liposuction specimens', Methods Mol Biol, 2008. 449: 69–79.
[28] Gottschling, S., et al., 'Mesenchymal stem cells in non-small cell lung cancer–different from others? Insights from comparative molecular and functional analyses', Lung Cancer, 2013. **80**(1): 19–29.
[29] Qiao, L., et al., 'Dkk-1 secreted by mesenchymal stem cells inhibits growth of breast cancer cells via depression of Wnt signalling', Cancer Lett, 2008. **269**(1): 67–77.
[30] Ho, I.A., et al., 'Human bone marrow-derived mesenchymal stem cells suppress human glioma growth through inhibition of angiogenesis', Stem Cells, 2013. **31**(1): 146–55.
[31] Loebinger, M.R., et al., 'Mesenchymal stem cell delivery of TRAIL can eliminate metastatic cancer', Cancer Res, 2009. **69**(10): 4134–42.
[32] Khakoo, A.Y., et al., 'Human mesenchymal stem cells exert potent antitumorigenic effects in a model of Kaposi's sarcoma', J Exp Med, 2006. **203**(5): 1235–47.

[33] Dasari, V.R., et al., 'Upregulation of PTEN in glioma cells by cord blood mesenchymal stem cells inhibits migration via downregulation of the PI3K/Akt pathway', PLoS One, 2010. 5(4): e10350.
[34] Stupp, R., et al., 'Radiotherapy plus concomitant and adjuvant temozolomide for glioblastoma', N Engl J Med, 2005. **352**(10): 987–96.
[35] Nakamura, K., et al., 'Antitumor effect of genetically engineered mesenchymal stem cells in a rat glioma model', Gene Ther, 2004. **11**(14): 1155–64.
[36] Nakamizo, A., et al., 'Human bone marrow-derived mesenchymal stem cells in the treatment of gliomas', Cancer Res, 2005. **65**(8): 3307–18.
[37] Bexell, D., et al., 'Bone marrow multipotent mesenchymal stroma cells act as pericyte-like migratory vehicles in experimental gliomas', Mol Ther, 2009. **17**(1): 183–90.
[38] Munoz, J.L., et al., 'Delivery of Functional Anti-miR-9 by Mesenchymal Stem Cell-derived Exosomes to Glioblastoma Multiforme Cells Conferred Chemosensitivity', Mol Ther Nucleic Acids, 2013. 2: e126.
[39] Bexell, D., S. Scheding, and J. Bengzon, 'Toward brain tumor gene therapy using multipotent mesenchymal stromal cell vectors', Mol Ther, 2010. **18**(6): 1067–75.

3

Prospective Technologies for Cardiac Repair

Francesca Pagliari[1,3], Luciana Carosella[2] and Paolo Di Nardo[3,4]

[1]Physical Science and Engineering Division, King Abdullah University of Science and Technology, Jeddah, Saudi Arabia
[2]Institute of Internal Medicine and Geriatrics, Catholic University of the Sacred Heart, Rome, Italy
[3]Laboratory of Molecular and Cellular Cardiology (LCMC), Department of Clinical Sciences and Translational Medicine, University of Rome "Tor Vergata", Rome, Italy
[4]Center for Regenerative Medicine, University of Rome Tor Vergata, Italy

Abstract

Cardiac diseases represent the major cause of death worldwide. Pharmacological treatments, although very sophisticated, are not able to definitively cure cardiac diseases. Furthermore, heart transplantation has shown to be efficient, but unsustainable because of donor shortage and extremely high costs of surgery and patient follow up. Finally, cell therapy applied to the injured myocardium has demonstrated to be inadequate to integrate a sufficient number of efficient contractile cells into the cardiac architecture. Considering the further expansion of cardiac diseases related to the explosive extension of longevity, it is urgent to formulate safe and cost-effective novel strategies to treat cardiac patients, without increasing the economic and social burden on public and private insurances as well as on families. Among others, the "selective repair" of the damaged region of a organ appears as the most reliable approach in the near future. Indeed, recent evidences have suggested that adult progenitor cells can be used to fabricate *ex vivo* engineered cardiac tissue to be implanted into the injured myocardium. However, novel materials and procedures to fabricate bio-compatible scaffolds are necessary to cope with the peculiar heart microenvironment and functional characteristics. In principle, engineered tissues can be fabricated using biocompatible polymeric scaffolds that remain embedded in the engineered tissue or, alternatively, the scaffold can be stuck on the petri bottom and the new tissue fabricated on, and not around, it. In the latter case, the engineered tissue will be scaffoldless. The current limitation of both technologies is that the scaffold is intended as a mere cell support. Instead, the scaffold must be active part in the array of biological signals governing the formation of a new tissue. This issue is very crucial in the specific case of engineered cardiac tissues that must repeat the native architecture and function. Indeed, preliminary results have shown that specifically manipulated biomaterials can be used to fabricate scaffolds inherently able to deliver signals sensed as "biologically relevant" by cells. The manipulation of the scaffold topology and nanostructure or the use of appropriate composite materials can allow to differentiate stem cells towards the cardiac phenotype in an architectural context very similar to the native one. This new class of scaffolds are very potent in addressing the cell phenotype when fine tuned in respect to the culture medium.

Alternatively, human progenitor cells, possibly isolated from the heart of the same patient candidate to receive the cell treatment, can be used to fabricate scaffoldless tissue sheets. When leant on the heart surface used as a scaffold, the scaffoldless tissue sheets release the embedded progenitor cells that easily migrate into the myocardium differentiating in cardiomyocytes and integrating in the tissue architecture, as demonstrated by the proper connections established between the graft and host cells.

Keywords: Cardiovascular disease, Tissue engineering, Progenitor cells, Additive manufacturing, Bioprinting.

3.1 Medicine Changing Needs

Medicine is undergoing an epochal revolution determined by the expanding ageing and, thus, sickening population, as never before in mankind history. The amplified awareness about the aging-related diseases (myocardial infarction, stroke, diabetes, cancer, etc.) and the advancements in biomedical research together with the increased wealth of industrialized and developing countries have generated great expectation about the possibility of developing very sophisticated treatments for their definitive cure while creating an equal access to the most advanced diagnostic and therapeutic procedures by all individuals independently of the geographic and economic conditions. This vision, besides its ethical and humanitarian relevance, involves very huge economical issues;

in fact, public and private insurances can hardly sustain present and future burden of degenerative diseases.

In this context, particular attention must be paid to cardiovascular diseases (CVD) that represent a growing health and socio-economic burden in most countries around the world [1, 2]. According to WHO, cardiovascular diseases are the most important cause of death and disability worldwide causing 24 million deceases/year by the end of 2030. WHO estimates that low- and middle-income countries are disproportionately affected: over 80% of global CVD deaths occur there proving that CVD is no longer a "rich white man's disease". According to the American Heart Association and the National Heart, Lung and Blood Institute, the staggering costs of treatments for CVD in the USA, including healthcare expenditures and lost productivity due to deaths and disability, were more than 500 billion USD in 2010. In the EU, CVD cause over 1.9 million deaths [3] and determine a total estimated annual cost of 169 billion EUR, with healthcare accounting for 62% of costs, which accounts for 10% of the healthcare expenditure across the Union. Informal care of patients and productivity losses exceeded another 115 billion [4]. However, the burden of CVD should not be measured by deaths alone; the cost in terms of human suffering and lost lives is incalculable. In this scenario, the available treatment options prolong the life span of cardiac patients, but these treatment modalities are not able to provide permanent solution for CVD. Furthermore, heart transplantation as well as implantable artificial heart pumps are available for few patients, only.

The lesson to be drawn from these alarming facts and figures is that the prevention and cure of cardiovascular diseases is not only an important medical necessity, but is also a social and economic imperative for the world sustainable growth and development [5]. Unfortunately, the development of novel efficient drugs requires impressive investments and no patents for major cardiovascular drugs have been registered in the last two decades. Also heart transplantation is restricted to few patients because of very high costs, organ shortage and possible immune rejection. Indeed, an insufficient number of heart transplantations is performed in North America and Europe only, while developing countries cannot afford the costs and, thus, their patients are excluded from this treatment option [6]. Renovated hopes have been raised by the evidence that stem cell therapy displays the potential to regenerate and repair the heart after injury [7, 8]. However, in spite of extensive investments and intensive investigations [9, 10], clinical trials have shown that only marginal benefits on heart function is induced by stem cell therapy in cardiac patients [11]. In fact, protocols and technologies so far used in cardiac cell therapy are rather rudimentary and do not consider the complexity of the myocardial tissue [12] that is a heterogeneous, anysotropic, viscoelastic system made of inert materials and a multiplicity of cell types (Figures 3.1 and 3.2). Cell therapy based on stem cells suspended in a culture medium and injected into the myocardium display major drawbacks, such as: (a) lack of a consensus on which cells types should be used; (b) lack of control of the potency of these cells; (c) poor cell survival and engraftment; (d) uncertain prevention of uncontrolled differentiation; (e) need for mechanistic understanding of cellular function in the therapeutic setting; (f) inexistence of instructive biomaterials that can promote and induce cell survival, migration and remodeling of the cardiac matrix; (g) lack of strategies to control the environment of the injured tissue to make it more suitable for the graft homing.

Recent advancements in cell biology and biomaterials research as well as the evolution of concepts and vision have indicated that Tissue Engineering could benefit the progress of regenerative medicine and its clinical application. Tissue Engineering is an emerging interdisciplinary field that aims at fabricating on the bench portions of biological tissues that can be used not only

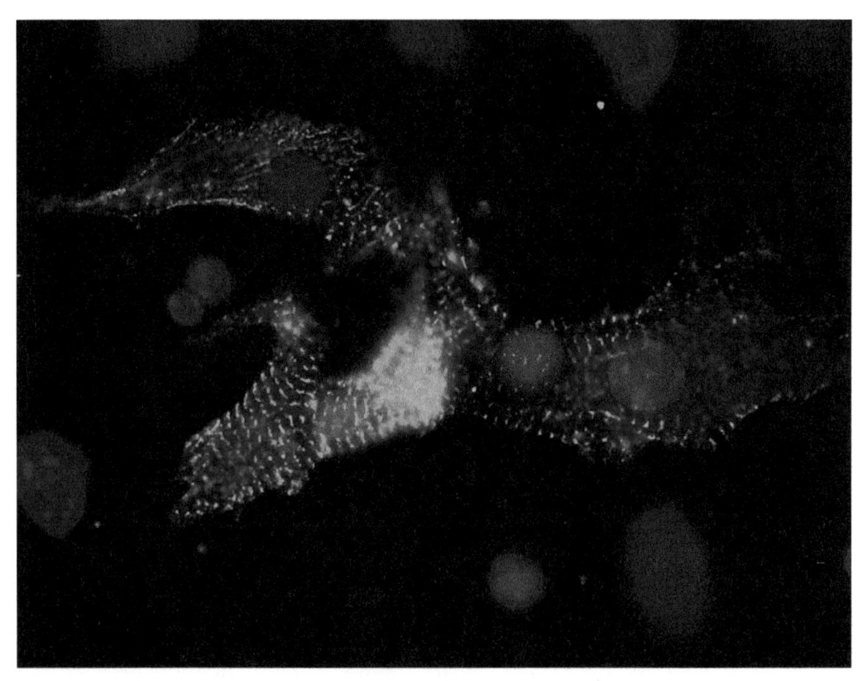

Figure 3.1 Mouse neonatal cardiomyocytes stained with α-sarcomeric actinin in green and nuclei in blue. Magnification 60×.

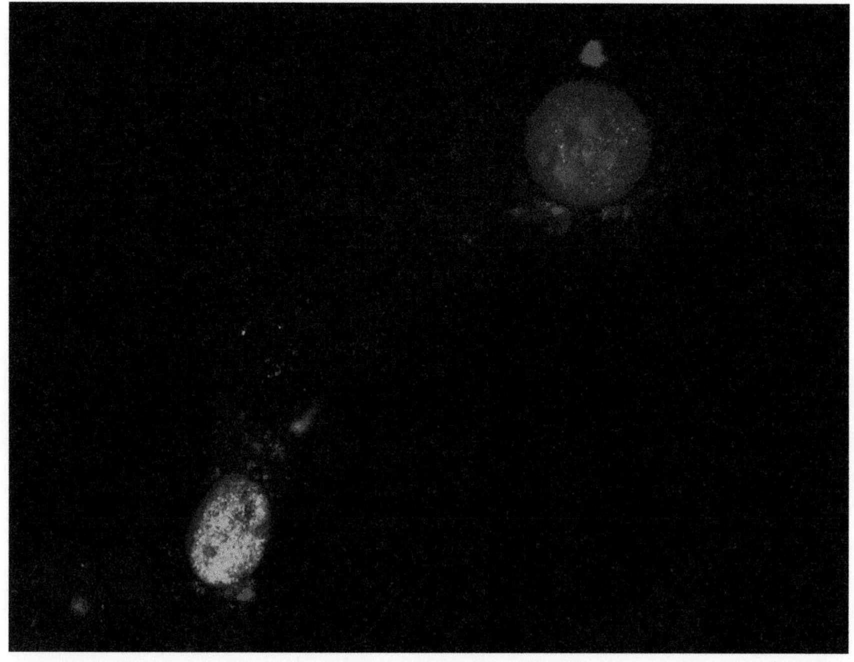

Figure 3.2 Cardiomyogenic differentiation of Vybrant red labeled-human cardiac progenitor cells (hCPCs) co-cultured with mouse neonatal cardiomyocytes. After 1 week, hCPCs show an up-regulation of GATA-4 (green) inside the nuclei. Magnification 60×.

to repair organs suffering from degenerative diseases [13], but also as controllable 3D models to study cell development and drug effects. The availability of human biological tissues will dramatically reduce the demand for whole heart (as well as other organs) transplantations, since only the damaged portions will be repaired overcoming organ shortage.

3.2 Additive Technologies in Tissue Engineering

Manufacturing engineered tissues implies the use of three components: functional scaffolds, cells and an appropriate environment. The fabrication of functional scaffolds is a very complex endeavor that entails the integration of the knowledge so far accumulated in different international laboratories in a multidisciplinary effort to finalize a very intricate procedure of manufacturing and implantation. This also implies that novel expertise must be created for the sake of patients and industries.

To fabricate engineered myocardial tissues, the first and most important achievement for the near future is to design and fabricate a scaffold very closely mimicking the ECM [14]. Innumerable scaffolds for myocardial tissue have been so far designed and experimentally tested, but none of them has demonstrated to be technologically ready for the clinical setting. The next generation of scaffolds must definitively allow cell growth and tissue organization thanks to a controlled architecture characterized by variations of the internal porosity with consequent enhanced control of interconnected channel networks to favor nutrient delivery, waste removal, exclusion of materials or cells, protein transport, and cell migration [13]. In addition, the scaffold must control cell fate releasing biochemical signals delivered by biomolecules (IGF, EGF, IL-1, IL-10, HGF, etc.). So far, biomolecules have been added in the scaffold starting solution with a resulting homogenous distribution into the scaffold itself. These results are quite far from the natural distribution of bioactive molecules that are distributed on the basis of very finely organized gradients [15]. Finally, scaffolds must release physical signals by itself (stiffness, tessellation, topology, etc.) and by incorporated micro/nanosystems able to generate adequate physical stimuli (e.g., electric field).

Different cell types grown on a multitude of scaffolds made of different materials have been so far investigated in order to fabricate strips of myocardial tissue [16–18]. Nevertheless, cell seeded scaffolds encounter host immune response, mechanical mismatch with the surrounding tissue, difficulty in uniformly integrate a high number of cells and limitations in incorporating multiple cell types with positional specificity [19]. Scaffoldless cell sheets [17] have also been manufactured (Figure 3.3), but protocols appear not yet reliable to allow clinical applications. Besides biological and biomaterial issues, manufacturing biological tissues requires sophisticated, rapid, extremely accurate and scalable techniques that are not manually operable. In this respect, a quantum leap has been made when it has been realized that additive manufacturing (3D printing) matches these requirements [20]. 3D printing allows to very precisely fabricate scaffolds layer-by-layer. A particular extension of additive technologies focused on living materials (bioprinting) permits to combine (i) different cell types, (ii) polymeric gels to mimic the extracellular matrix, (iii) immuno-suppressive soluble factors to prevent rejection and (iv) biochemical substances to control the behavior [21]. The ultimate goal is to reproduce complex heterogeneous immuno-privileged/biocompatible biological tissues, either by positioning different cell types in desired locations or by inducing progenitor cells to differentiate

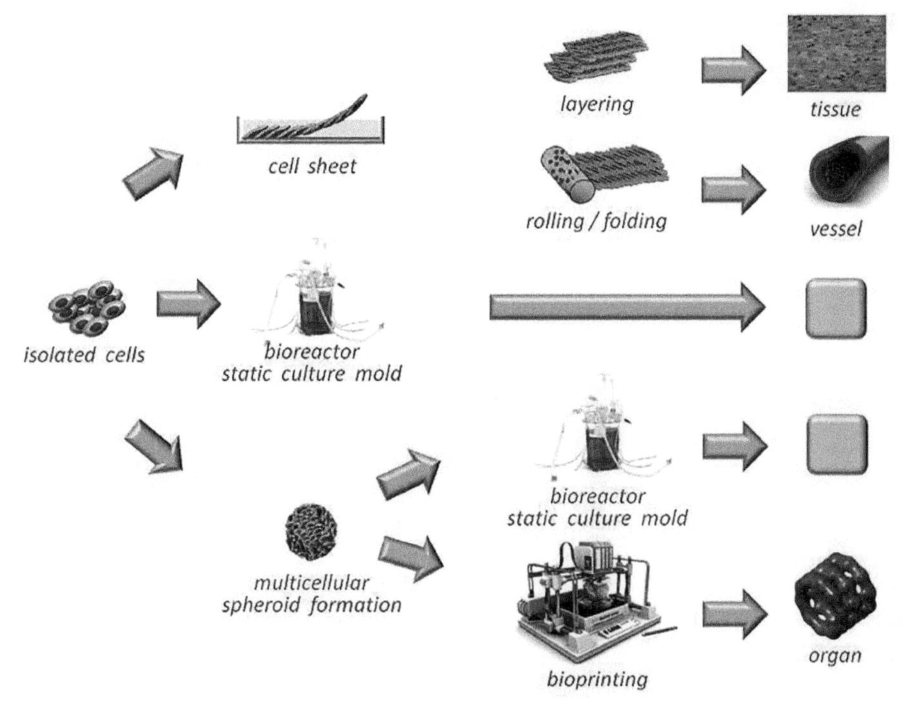

Figure 3.3 Scaffoldless Biofabrication. Engineered tissues can be fabricated in the absence of polymeric structures (scaffolds) supporting cell growth. The figure summarizes the different strategies under experimentation. Techniques to manufacture cell sheets (*upper figure*) are the most simple and generate engineered tissues made of a single cell population. Instead, 3D Bioprinting can allow to closely mimic complex tissue architectures.

into the desired cell types in the context of a specific bio-architecture. In this respect, environmental parameters (such as pH, pressure and geometry of the surrounding space) greatly affect the behavior and the differentiation of stem cells, slowing down or speeding up its dynamics and/or addressing toward different types of differentiation.

The complexity of the tissue to be fabricated cannot be reproduced by manual procedures, as in conventional laboratory techniques. It requires that novel materials, technologies and protocols are exploited or invented through a long-term process actuated by merging the quantum of knowledge resident in different disciplines and international laboratories. Ambitious ideas must be exploited in joint continued collaborative efforts in which the risk and the possible failure are central factor of innovation. In this context, materials and procedures must be strictly standardized involving the knowledge accumulated in a multiplicity of fields, such as biology, medicine, mathematics, ICT, material science and engineering. An example of this approach is the Additive Manufacturing in which a CAD software governs the nozzles of a 3D printer modified to deposit layer-by-layer and to pattern the biocompatible polymeric gels (biopaper) on which small amounts of cells (bioink) are positioned on the basis of a specific architecture. In this process, the biopaper plays a pivotal role. In fact, stem cells are prone to adopt the final phenotype only when cultured in strictly controlled conditions characterized by a critical array of chemical, biochemical and physical factors, emulating the ECM environment of the original tissue. When adequately manipulated and designed, the biopaper (also without embedded biological molecules) can release signals perceived as biologically relevant by cells, as otherwise demonstrated in studies on cardiomyocyte differentiation [22, 23]. Alternative bioprinting protocols allow fabricating scaffoldless bio-tissues. Cells stick and move together in clumps with liquid-like properties during embryogenesis [24, 25]. A bio-mimetic approach to the fabrication of engineered tissues inspired by the mechanisms presiding over cell self-assembling in the absence of scaffolds during embryogenesis is actively investigated. Genetic and physical interplay drives cells self-assembling in microspheroids that constitute the tissue building blocks [26]. Micro-spheroids are fluidic-like and their fusion is driven by surface tension forces and by the "differential adhesion hypothesis" which postulates that cells of diverse types adhere to each other with different strength due to either quantitative or qualitative differences in cell surface adhesion molecules [27]. A mixed population of differentially adhesive cells evolves in a compartmentalized system in which the less adhesive surround the self-aggregated most adhesive cells. This process is also participated by the cellular tensile forces generated by acto-myosin-dependent cell cortex tension. In a subsequent step, the progressive accumulation of self-produced extracellular matrix restricts cell motility and enhances tissue cohesion modulating tissue fusion processes [28]. The synergistic interaction of self-assembling spheroids and self-assembling matrix material ultimately leads to hierarchically ordered structures inducing the evolution of the cell system from an initial to a more stable state [29–31]. However, in the absence of a solid scaffold, engineered self-assembled tissues must undergo a rapid fluid-solid transition to preserve their shape, composition and integrity.

In a typical bioprinting approach, mechanical extruders place multi-cellular aggregates of definite composition (bioink particles) according to a computer-generated template together with hydrogel (biopaper) constituting the supporting environment. The post-printing fusion of bioink particles generates organoids taking advantage by early developmental mechanisms such as cell sorting and fusion. Taken together, printing-based tissue engineering technology allows (i) producing fully biological (scaffold-free) small diameter tissues; (ii) it is based on natural shape-forming (i.e. morphogenetic) processes, that are present during normal development; (iii) it can provide organoids of complex topology (i.e, branching tubes); (iv) it is scalable and compatible with methods of rapid prototyping. The ultimate goal of this technology is to generate protocols to instruct (rather than to use) cells to fabricate tissues through a highly engineered artificial environmental milieu. Indeed, the full biological potential of stem cells (*in vitro* or *in vivo*) can be deployed only in an environment mimicking native development. Bioprinting technology holds promise to allow reproducing this complex environment even if an intense investigation activity must be undertaken to fine-tune all the innumerable details that can guarantee a successful output.

References

[1] C.J. Murray, A.D. Lopez. Global mortality, disability, and the contribution of risk factors: Global Burden of Disease Study. Lancet 349(9063):1436–42, 1997.

[2] R. Kahn, R.M. Robertson, R. Smith, D. Eddy, 'The impact of prevention on reducing the burden of cardiovascular disease' Circulation, 118, 576–585, 2008.

[3] S. Petersen, V. Peto, M. Rayner, J. Leal, R. Luengo-Fernandez, A. Gray. European cardiovascular disease statistics, British Heart Foundation.

[4] J. Leal, R. Luengo-Fernández, A. Gray, S. Petersen, M. Rayner. Economic burden of cardiovascular diseases in the enlarged European Union. Eur Heart J. 27: 1610–1619, 2006.

[5] W.S. Weintraub, S.R. Daniels, L.E. Burke, B.A. Franklin, D.C. Jr Goff, L.L. Hayman, D. Lloyd-Jones, D.K. Pandey, E.J. Sanchez, A.P. Schram, L.P. Whitsel; American Heart Association Advocacy Coordinating Committee; Council on Cardiovascular Disease in the Young; Council on the Kidney in Cardiovascular Disease; Council on Epidemiology and Prevention; Council on Cardiovascular Nursing; Council on Arteriosclerosis; Thrombosis and Vascular Biology; Council on Clinical Cardiology, Stroke Council 'Value of primordial and primary prevention for cardiovascular disease: a policy statement from the American Heart Association' Circulation, 124, 967–990, 2011.

[6] C. Benden, L.B. Edwards, A.Y. Kucheryavaya, J.D. Christie, A.I. Dipchand, F. Dobbels, R. Kirk, A.O. Rahmel, J. Stehlik, M.I. Hertz MI, International Society of Heart and Lung Transplantation, 'The Registry of the International Society for Heart and Lung Transplantation: fifteenth pediatric lung and heart-lung transplantation report—2012' J. Heart Lung Transplant., 31, 1087–1095, 2012.

[7] K. Urbanek, D. Torella, F. Sheikh, A. De Angelis, D. Nurzynska, F. Silvestri, C.A. Beltrami, R. Bussani, A.P. Beltrami, F. Quaini, R. Bolli, A. Leri, J. Kajstura, P. Anversa, 'Myocardial Regeneration by Activation of Multipotent Cardiac Stem Cells in Ischemic Heart Failure' Proc. Natl. Acad. Sci. USA., 102, 8692–8697, 2005.

[8] O. Bergmann, R.D. Bhardwaj, S. Bernard, S. Zdunek, F. Barnabé-Heider, S. Walsh, J. Zupicich, K. Alkass, B.A. Buchholz, H. Druid, S. Jovinge, J. Frisén, 'Evidence for Cardiomyocyte Renewal in Humans' Science, 324, 98–102, 2009.

[9] R. Bolli, A.R. Chugh, D. D'Amario, J.H. Loughran, M.F. Stoddard, S. Ikram, G.M. Beache, S.G. Wagner, A. Leri, T. Hosoda, F. Sanada, J.B. Elmore, P. Goichberg, D. Cappetta, N.K. Solankhi, I. Fahsah, D.G. Rokosh, M.S. Slaughter, J. Kajstura, P. Anversa, 'Cardiac stem cells in patients with ischaemic cardiomyopathy (SCIPIO): initial results of a randomised phase 1 trial' Lancet, 378, 1847–1857, 2011.

[10] K. Malliaras, R.R. Makkar, R.R. Smith, K. Cheng, E. Wu, R.O. Bonow, L. Marbán, A. Mendizabal, E. Cingolani, P.V. Johnston, G. Gerstenblith, K.H. Schuleri, A.C. Lardo, E. Marbán, 'Intracoronary cardiosphere-derived cells after myocardial infarction: evidence of therapeutic regeneration in the final 1-year results of the CADUCEUS trial (CArdiosphere-Derived aUtologous stem CElls to reverse ventricUlar dySfunction)' J. Am. Coll. Cardiol., 63, 110–122, 2014.

[11] A. Abbott, 'Doubts over heart stem-cell therapy' Nature, 509, 15–16, 2014, Erratum in: Nature, 509, 272, 2014.

[12] P. Di Nardo, G. Forte, A. Ahluwalia, M. Minieri. Cardiac progenitor cells: potency and control. J Cell Physiol 224: 590–600, 2010.

[13] R. Langer, J.P. Vacanti, 'Tissue engineering' Science, 260, 920–926, 1993.

[14] T. Dvir, B.P. Timko, D.S. Kohane, R. Langer, 'Nanotechnological strategies for engineering complex tissues' Nat. Nanotechnol., 6, 13–22, 2011.

[15] E.M. Pera, H. Acosta, N. Gouignard, M. Climent, I. Arregi, 'Active signals, gradient formation and regional specificity in neural induction' Exp. Cell. Res., 321, 25–31, 2014.

[16] G. Forte, S. Pietronave, G. Nardone, A. Zamperone, E. Magnani, S. Pagliari, F. Pagliari, C. Giacinti, C. Nicoletti, A. Musaró, M. Rinaldi, M. Ribezzo, C. Comoglio, E. Traversa, T. Okano, M. Minieri, M. Prat, P. Di Nardo. Human cardiac progenitor cell grafts as unrestricted source of supernumerary cardiac cells in healthy murine hearts. Stem Cells. 29: 2051–2061, 2011.

[17] M. Kawamura, S. Miyagawa, K. Miki, A. Saito, S. Fukushima, T. Higuchi, T. Kawamura, T. Kuratani, T. Daimon, T. Shimizu, T. Okano, Y. Sawa, 'Feasibility, safety, and therapeutic efficacy of human induced pluripotent stem cell-derived cardiomyocyte sheets in a porcine ischemic cardiomyopathy model' Circulation, 126, S29–37, 2012.

[18] Y.S. Choi, K. Matsuda, G.J. Dusting, W.A. Morrison, R.J. Dilley, 'Engineering cardiac tissue in vivo from human adipose-derived stem cells' Biomaterials, 31, 2236–2242, 2010.

[19] G. Vunjak-Novakovic, N. Tandon, A. Godier, R. Maidhof, A. Marsano, T.P. Martens, M. Radisic, 'Challenges in cardiac tissue engineering' Tissue Eng. Part B Rev., 16, 169–187, 2010.

[20] B. Derby. Printing and prototyping of tissues and scaffolds. Science 338: 921–926, 2012.

[21] S.V. Murphy, A. Atala, '3D bioprinting of tissues and organs' Nat. Biotechnol., 32, 773–785, 2014.

[22] S. Pagliari, A.C. Vilela-Silva, G. Forte, F. Pagliari, C. Mandoli, G. Vozzi, S. Pietronave, M. Prat, S. Licoccia, A. Ahluwalia, E. Traversa, M. Minieri, P. Di Nardo, 'Cooperation of biological and mechanical signals in cardiac progenitor cell differentiation' Adv. Mater., 23, 514–518, 2011.

[23] F. Pagliari, C. Mandoli, G. Forte, E. Magnani, S. Pagliari, G. Nardone, S. Licoccia, M. Minieri, P. Di Nardo, E. Traversa. Cerium oxide nanoparticles protect cardiac progenitor cells from oxidative stress. *ACS Nano* 6: 3767–3775, 2012.

[24] R.A. Foty, C.M. Pfleger, G. Forgacs, M.S. Steinberg. Surface tensions of embryonic tissues predict their mutual envelopment behavior. *Development* 122: 1611–1620, 1996.

[25] G. Forgacs, R.A. Foty, Y. Shafrir, M.S. Steinberg, 'Viscoelastic properties of living embryonic tissues: a quantitative study' Biophys J., 74, 2227–2234, 1998.

[26] V. Mironov, R.P. Visconti, V. Kasyanov, G. Forgacs, C.J. Drake, R.R. Markwald. Organ printing: tissue spheroids as building blocks. Biomaterials 30: 2164–2174, 2009.

[27] R.A. Foty, M.S. Steinberg. The differential adhesion hypothesis: a direct evaluation. Dev. Biol., 278, 255–263, 2005.

[28] M. McCune, A. Shafiee, G. Forgacs, I. Kosztin. Predictive modeling of post bioprinting structure formation. Soft Matter 10: 1790–800, 2014.

[29] E. D. F. Ker et al. Bioprinting of growth factors onto aligned sub-micron fibrous scaffolds for simultaneous control of cell differentiation and alignment. *Biomaterials* 32: 8097–107, 2011.

[30] K. Jakab, C. Norotte, B. Damon, F. Marga, A. Neagu, C.L. Besch-Williford, A. Kachurin, K.H. Church, H. Park, V. Mironov, R. Markwald, G. Vunjak-Novakovic, G. Forgacs. Tissue engineering by self-assembly of cells printed into topologically defined structures. *Tissue Eng Part A* 14: 413–421, 2008.

[31] J.A. Phillippi, E. Miller, L. Weiss, J. Huard, A. Waggoner, P. Campbell. Microenvironments engineered by inkjet bioprinting spatially direct adult stem cells toward muscle- and bone-like subpopulations. *Stem Cells* 26: 127–134, 2008.

4

Does Inter-Individual Heterogeneity in the Normal Breast Corrupt Cancer Stem Cell and/or Cancer-Specific Signaling Characterization?

Harikrishna Nakshatri

Departments of Surgery, Biochemistry and Molecular Biology,
Indiana University Simon Cancer Center,
Indiana University School of Medicine,
Indianapolis, Indiana, USA

Abstract

Several studies have described cell surface markers that phenotypically define stem-progenitor-mature cell hierarchy in the normal breast. The same markers have been used to identify subpopulation of cancer cells with enhanced tumor initiating capacity. These subpopulations of cells, also called cancer stem cells (CSCs), have been the focus of intense research for the last few years. Identifying and characterizing cancer-specific differences in CSCs from their normal counter part is not trivial due to non-availability of replenishable source of primary normal and CSCs to perform functional assays. Moreover, recent discovery of widespread genetic variation in humans leading to functional transcriptome diversity makes the task of defining "global normal" very difficult. Thus, "normal" breast epithelial hierarchy and corresponding gene expression profiles have to be defined at individual patient level for comparison with cancer. Recent advances in human mammary epithelial cell reprogramming growth conditions and single cell genome analyses should overcome these limitations and enable characterization of "normal" and "tumor" at individual levels. By propagating cells from core breast biopsies of healthy donors, tumors and adjacent normal followed by flow cytometry analysis, we have recently observed remarkable inter-individual phenotypic heterogeneity in normal breast stem, luminal progenitor, and mature cell numbers and possibly epithelial cell plasticity. Comparison of adjacent normal and tumor from the same patient showed distinct differences in differentiation status between normal and tumor. This observation has important implications as cancer-specific defect in differentiation alone could account for the majority of gene expression differences observed between cancer and normal cells. In addition, most of the differentially expressed genes including genes with highest expression difference, which are often considered for functional studies or as biomarkers of cancer cell behavior, are not causally linked to cancer. Collectively, inter-individual heterogeneity in the normal breast, the differences in the differentiation status between normal and tumor of the same patient, and differences in epithelial cellularity between normal and tumors used for gene expression studies may be the reasons for discrepancy in the literature with respect to gene expression based prognostic signatures, cancer-specific signaling pathway alterations, and CSC characterization. As a way forward, we propose that the magnitude of tumor heterogeneity and CSC phenotype is the product of individual's epithelial cell plasticity and cancer-specific mutation. In addition, we need to characterize normal and tumor on an individual basis for clear understanding of pathobiology of tumors.

Keywords: Normal breast, Epithelial hierarchy, Heterogeneity, Breast cancer, Cancer stem cells.

4.1 Introduction

Recent advances in genomics have shown profound inter-individual functional diversity in transcriptome due to genetic regulatory variations [1]. In a study involving non-transformed fibroblasts from 62 unrelated individuals, Wagner et al. found significant inter-individual differences in the expression levels of 9,493 out of 16,952 genes with strongest differences in the expression levels of a subset of developmentally regulated Hox gene cluster [2]. This breakthrough should force us to redefine "normal" and reassess cancer-specific variation from normal variation in gene expression. The integrative cluster classification of breast cancer study addressed this issue partially, where the impact of inherited copy number variations and single nucleotide variations were taken into consideration to derive cancer-specific transcriptome [3]. However, the problem still persists when one defines "normal" without any consideration to ethnic differences in the normal tissues used as controls. In normal breast, menstrual cycle at the time of tissue collection adds another variability to transcriptome [4]. Along this line, lack of reproducibility of gene expression signatures or biomarkers with prognostic significance may in part be due to genetic regulatory variation in "normal". In addition, methods to develop targeted therapies need to be reevaluated because some of signaling network considered to be active based on comparison of cancer transcriptome with "global normal" breast transcriptome may be misleading. Indeed, analysis using newer computational tools suggest that driver mutations required for oncogenesis are relatively small suggesting that few of the previously described cancer-specific aberrations in gene expression are not causally linked to cancer [5]. Instead, most of the documented differences in gene expression between normal and cancer are likely due to inter-individual heterogeneity in normal breast transcriptome.

4.2 Defining Normal Breast Hierarchy

Although it is ideal to document inter-individual heterogeneity of the normal breast at genomic levels, it is not is easily achievable. A simplest assay would be to characterize normal cells for cell surface markers that have previously been used for stem, progenitor and mature cell identification. Several recent reviews by others and us provide a list of such markers [6–8]. For example, CD49f+/EpCAM−, CD49f+/EpCAM+, and CD49f−/EpCAM+ cells correspond to stem, luminal progenitor, and mature/differentiated cells of breast, respectively [8]. EpCAM+/CD49f+CD10+ are bipotent cells, EpCAM+/CD49f+/MUC1+ cells are luminal-restricted colony forming cells, EpCAM+/CD49f−/MUC1+ cells are mature luminal cells and EpCAM+/CD49f−/MUC1−/CD10+ cells are differentiated myoepithelial cells of the human breast [9]. Basal cells of the breast express CD271 [10]. Basal cells of the normal breast also demonstrate CD44+/CD24− phenotype [11]. CD73+/CD90− cells correspond to rare cells in the breast that exhibit extensive lineage plasticity [12]. Limited gene expression studies using purified subpopulation of cells from normal breast have demonstrated significant gene expression differences between these populations. For example, ~2000 genes are differentially expressed between bipotent luminal and mature luminal cells [9]. Similar differences in gene expression between CD44+/CD24− and CD44−/CD24+ cells have been observed [11]. Using non-transformed MCF-10A breast epithelial cell line, we had demonstrated differential expression of >2000 genes between CD44+/CD24− and CD44−/CD24+ epithelial cells [13]. These subpopulation-specific differences also extend to microRNAs; basal, luminal progenitor and mature cells of normal breast express different set of microRNAs [14].

Although the above-mentioned studies documented the presence of different populations of cells in the normal breast, this knowledge has not been utilized extensively to characterize tumor-specific gene expression. The following limitations may have had a negative impact on cancer-specific transcriptome analyses. 1) Most of the "normal" tissues used for comparative gene expression studies were derived from either reduction mammoplasty or contiguous with the tumor; 2) There are no replenishable primary cells to determine the function of genes uniquely expressed in a subpopulation of cells; 3) Since gene expression analyses were done using flow sorted cells directly [9, 11], in which gene expression could still be under the influence of microenvironment, it is difficult to determine intrinsic gene expression pattern in subpopulation of cells; 4) Because of the nature of normal tissues utilized, inter-individual heterogeneity in number of phenotypically defined subpopulations in the normal breast due to age, body mass index, ethnicity, parity, breast feeding, use of birth control pills, menstrual cycle at the time of tissue collection, menopausal status, or age at menarche corrupts the definition of normal gene expression pattern in breast. One would expect expression changes in >2000 genes simply due to differences in progenitor to mature cell ratio between healthy individuals. Attempting to address these issues is not easy but achievable as explained below.

4.3 Need for *In Vitro* Assays to Document Inter-Individual Heterogeneity in the Normal Breast

Over the years, a series of breast epithelial cell lines have been generated to study the role of specific oncogenes, transcription factors, epithelial to mesenchymal transition (EMT), growth factors and cytokines in breast cancer progression. However, most of these cell lines display basal cell features [15]. Using different media composition, Weinberg's group was able to generate two distinct subtypes of normal cells (with basal and luminal features) but neither contained estrogen receptor alpha (ERα)-positive cells [16]. Thus the currently available model systems cannot document or study inter-individual heterogeneity. Lack of the model systems has also prevented any mechanistic studies on tumor initiating events responsible for specific subtypes of breast cancer. For example, based on microarray analysis, breast cancer is classified into five intrinsic subtypes; luminal A, luminal B, normal-like/claudin-low, Her2+ and basal type [17]. Luminal A and luminal B express ERα.

ERα-positive breast cancer represents ~70% of breast cancer cases and is a major clinical problem. Although ERα-positive breast cancers are generally thought to be less aggressive, luminal B ERα-positive breast cancers show poor outcome, almost similar to basal-type breast cancers [3, 17]. Similarly, integrative cluster analysis has also identified three ERα-positive integrative clusters with differing outcomes [3]. Mechanisms responsible for differential outcome in these ERα-positive subtypes can be revealed only when we know the ERα-signaling network in non-transformed breast epithelial cells.

In the normal mammary gland of human, rats, mice and cows, ERα-positive cells are heterogeneously located in the luminal compartment of the duct and rarely co-localize with proliferating cells [18]. ERα-positive tumor cells, in contrast, proliferate in response to estradiol (E2) treatment. In mouse models, autocrine activity of transforming growth factor beta (TGFβ) prevents ERα-positive cells from responding to E2 and impairment of TGFβ signaling is essential for ERα-positive cells to acquire E2-dependent proliferation [19]. However, similar studies in human breast epithelial cells have not been conducted due to the failure of currently available culture systems to support growth of non-transformed ERα-positive cells. Since the number of normal ERα-positive cells vary between individuals (5−20%), understanding inter-individual heterogeneity in E2-dependent autocrine and paracrine gene expression changes in the normal breast is essential for defining transcriptome in the normal breast.

Researchers have attempted to address the above issue by reintroducing ERα to ERα-negative cells. Paradoxically, introduced ERα inhibited growth instead of supporting E2-dependent proliferation [20]. Furthermore, there is evidence in the literature that ERα target genes are methylated in the absence of ERα [21]. Only way to activate endogenous ERα in these ERα-negative cells is to treat cells with DNA methyltransferase and histone deacetylase inhibitors, which have additional effects on the genome [22]. Therefore, resources need to be applied to develop a system that allows culturing of non-transformed ERα-positive cells.

Recently developed epithelial cell reprogramming assay will likely change the landscape of breast cancer research and will enable us to address several of these unmet needs stated above [23]. Indiana University houses Susan G Komen for the Cure normal breast tissue bank to which healthy donors donate breast core biopsies. This resource should eliminate the major concern regarding "normal" breast tissues that are being used as healthy controls in various gene expression studies including The Cancer Genome Atlas (TCGA) [24]. We have begun to address these issues by culturing core biopsies of healthy donors, high risk patients, adjacent normal and tumor cells from the same patient for s short duration and subjected these cells to phenotypic analysis using various cell surface markers and flow cytometry. We observed remarkable phenotypic heterogeneity in the normal breast among healthy donors irrespective of their age, menstrual cycle, body mass index, and parity [25]. The number of CD49f+/EpCAM−, CD49f+/EpCAM+, and CD49f−/EpCAM+ cells varied from individual to individual.

Tumor and adjacent normal cells from the same patient were phenotypically different indicating differences in differentiation status, which we expect to have an impact on gene expression pattern [25]. Furthermore, we did not find significant differences in the levels of CD44+/CD24− cells between adjacent normal and tumor cells of the same patient. Therefore, without characterizing normal breast of the same patient, it is difficult to conclude enrichment of CD44+/CD24− CSCs in a tumor of a patient. In contrast, we found enrichment of ALDEFLUOR-positive CSCs in tumors compared with adjacent normal depending on the cancer type. Since luminal progenitor or committed luminal cells are ALDEFLUOR-positive [26], the above observation indicates that cancers that originate from luminal cells and/or maintain luminal features are ALDEFLUOR-positive.

The ability to culture primary cells cultured under reprogramming condition may allow us to address another longstanding question related to ERα function in non-transformed cells. We expect a subpopulation of these cells to express ERα and consequently respond to E2, which will likely be different from ERα-positive breast cancer cells.

Epidemiologic studies have shown pregnancy protects against breast cancer [27]. A recent study has shown that CD44+p27+ positive cells with progenitor properties confer susceptibility to breast cancer and the number of CD44+p27+ cells in the breast decline with pregnancy [28]. As noted above, there can also be inter-individual heterogeneity in the number of p27+ stem cells and rate at which they decline after each pregnancy may help to assess breast cancer risk. Primary breast epithelial cells grown from nulliparous and parous women may allow risk assessment as well as to conduct mechanistic studies related to susceptibility of CD44+/p27+ cells to tumorigenesis.

4.4 Future Directions

Recent advances in next-generation sequencing, primary cell culturing, availability of truly normal breast tissues, and 3D culture techniques should enable major strides against breast cancer. There have been lots of efforts to understand tumor heterogeneity. However, heterogeneity in normal breast has not been the forefront of research efforts. Tumor heterogeneity in some cases could be a reflection of inherent properties of breast epithelial cells of the individual rater than due to cancer-specific genomic aberration. Distinguishing inherent heterogeneity from mutation-induced heterogeneity will have an impact on how tumors are characterized, cancer-specific signaling networks are identified, and treatment decisions are made.

List of Abbreviations:
CSC, Cancer Stem Cells
E2, estradiol
ERα, Estrogen Receptor alpha
TCGA, The Cancer Genome Atlas
TGFβ, Transforming Growth Factor Beta

Acknowledgements

The Susan G. Komen for the Cure supports research in Nakshatri's laboratory.

References

[1] Lappalainen T, Sammeth M, Friedlander MR, *et al.* Transcriptome and genome sequencing uncovers functional variation in humans. Nature 2013;501(7468):506–11.

[2] Wagner JR, Busche S, Ge B, *et al.* The relationship between DNA methylation, genetic and expression inter-individual variation in untransformed human fibroblasts. Genome Biol 2014;15(2):R37.

[3] Curtis C, Shah SP, Chin SF, *et al.* The genomic and transcriptomic architecture of 2,000 breast tumours reveals novel subgroups. Nature 2012;486(7403):346–52.

[4] Pardo I, Lillemoe HA, Blosser RJ, *et al.* Next-generation transcriptome sequencing of the premenopausal breast epithelium using specimens from a normal human breast tissue bank. Breast Cancer Res 2014;16(2):R26.

[5] Kandoth C, McLellan MD, Vandin F, *et al.* Mutational landscape and significance across 12 major cancer types. Nature 2013;502(7471): 333–9.

[6] Nakshatri H, Srour EF, Badve S. Breast cancer stem cells and intrinsic subtypes: controversies rage on. Curr Stem Cell Res Ther 2009;4(1): 50–60.

[7] Badve S, Nakshatri H. Breast-cancer stem cells-beyond semantics. Lancet Oncol 2012;13(1):e43–8.
[8] Visvader JE, Stingl J. Mammary stem cells and the differentiation hierarchy: current status and perspectives. Genes Dev 2014;28(11): 1143–1158.
[9] Raouf A, Zhao Y, To K, *et al.* Transcriptome analysis of the normal human mammary cell commitment and differentiation process. Cell Stem Cell 2008;3(1):109–18.
[10] Kim J, Villadsen R, Sorlie T, *et al.* Tumor initiating but differentiated luminal-like breast cancer cells are highly invasive in the absence of basal-like activity. Proc Natl Acad Sci USA 2012;109(16):6124–9.
[11] Shipitsin M, Campbell LL, Argani P, *et al.* Molecular definition of breast tumor heterogeneity. Cancer Cell 2007;11(3):259–73.
[12] Roy S, Gascard P, Dumont N, *et al.* Rare somatic cells from human breast tissue exhibit extensive lineage plasticity. Proc Natl Acad Sci USA 2013;110(12):4598–603.
[13] Bhat-Nakshatri P, Appaiah H, Ballas C, *et al.* SLUG/SNAI2 and tumor necrosis factor generate breast cells with CD44+/CD24− phenotype. BMC Cancer 2010;10:411.
[14] Pal B, Chen Y, Bert A, *et al.* Integration of microRNA signatures of distinct mammary epithelial cell types with their gene expression and epigenetic portraits. Breast Cancer Res 2015;17:85.
[15] Stampfer MR, Yaswen P. Culture models of human mammary epithelial cell transformation. J Mammary Gland Biol Neoplasia 2000;5(4): 365–78.
[16] Ince TA, Richardson AL, Bell GW, *et al.* Transformation of different human breast epithelial cell types leads to distinct tumor phenotypes. Cancer Cell 2007;12(2):160–70.
[17] Sorlie T, Perou CM, Tibshirani R, *et al.* Gene expression patterns of breast carcinomas distinguish tumor subclasses with clinical implications. Proc Natl Acad Sci USA 2001;98(19):10869–74.
[18] Grimm SL, Rosen JM. Stop! In the name of transforming growth factor-beta: keeping estrogen receptor-alpha-positive mammary epithelial cells from proliferating. Breast Cancer Res 2006;8(4):106.
[19] Ewan KB, Oketch-Rabah HA, Ravani SA, *et al.* Proliferation of estrogen receptor-alpha-positive mammary epithelial cells is restrained by transforming growth factor-beta1 in adult mice. Am J Pathol 2005;167(2):409–17.

[20] Jeng MH, Jiang SY, Jordan VC. Paradoxical regulation of estrogen-dependent growth factor gene expression in estrogen receptor (ER)-negative human breast cancer cells stably expressing ER. Cancer Lett 1994;82(2):123–8.

[21] Leu YW, Yan PS, Fan M, *et al.* Loss of estrogen receptor signaling triggers epigenetic silencing of downstream targets in breast cancer. Cancer Res 2004;64(22):8184–92

[22] Sharma D, Blum J, Yang X, *et al.* Release of methyl CpG binding proteins and histone deacetylase 1 from the Estrogen receptor alpha (ER) promoter upon reactivation in ER-negative human breast cancer cells. Mol. Endocrinol 2005;19(7):1740–51.

[23] Liu X, Ory V, Chapman S, *et al.* ROCK inhibitor and feeder cells induce the conditional reprogramming of epithelial cells. Am J Pathol 2012;180(2):599–607.

[24] Koboldt DC, Fulton RS, McLellan MD, *et al.* Comprehensive molecular portraits of human breast tumours. Nature 2012;490:61–70.

[25] Nakshatri H, Anjanappa M, Bhat-Nakshatri P. Ethnicity-Dependent and – Independent Heterogeneity in Healthy Normal Breast Hierarchy Impacts Tumor Characterization. Sci Rep 2015;5:13526.

[26] Lim E, Wu D, Pal B, *et al.* Transcriptome analyses of mouse and human mammary cell subpopulations reveal multiple conserved genes and pathways. Breast Cancer Res 2010;12(2):R21.

[27] Colditz GA, Rosner BA, Chen WY, *et al.* Risk factors for breast cancer according to estrogen and progesterone receptor status. J Natl Cancer Inst 2004;96(3):218–28.

[28] Choudhury S, Almendro V, Merino VF, *et al.* Molecular profiling of human mammary gland links breast cancer risk to a p27(+) cell population with progenitor characteristics. Cell Stem Cell 2013;13(1):117–30.

5

The Role of Physical Microenvironmental Cues on Myogenesis: Implications for Tissue Engineering of Skeletal Muscle

Cristian Pablo Pennisi[1], Stavros Papaioannou[1], John Rasmussen[2] and Vladimir Zachar[1]

[1]Laboratory for Stem Cell Research, Department of Health Science and Technology, Aalborg University, 9220 Aalborg, Denmark
[2]The AnyBody Research Group, Department of Mechanical and Manufacturing Engineering, Aalborg University, Aalborg, Denmark

Abstract

Skeletal muscle tissue engineering holds promise for the treatment of a variety of degenerative and traumatic muscle conditions. The ultimate goal of skeletal muscle tissue engineering is very challenging, since it needs to closely reproduce a highly complex structure consisting of muscle cells, connective tissue, blood vessels, and nerves. One of the fundamental prerequisites is to obtain cells arranged in parallel to efficiently produce force by contraction. Several studies have demonstrated that structural cues, such as topography and stiffness of the extracellular microenvironment, are crucial to control the organization and behavior of the cells. In addition to these passive signals, dynamic electrical and mechanical signals are also important for the assembly and maturation of skeletal muscle cells. The aim of this review is to present the current *in vitro* approaches used to investigate the influence of physical microenvironmental cues in skeletal myogenesis. These experimental approaches have been very valuable to answer fundamental biological questions and may help developing novel platforms for the fabrication of tissue-engineered muscle.

Keywords: Skeletal muscle, Tissue engineering, Microtopography, Skeletal myogenesis, Stiffness, Cyclic tensile strain, Uniaxial strain.

5.1 Introduction

Skeletal muscle comprises up to 40% of the adult human body weight, representing the largest tissue class in the body. Proper muscle function is fundamental for carrying out the voluntary movements of everyday life activities and for maintaining the metabolic homeostasis of the body [1]. Although adult muscle tissue possesses an exceptional capacity for regeneration, restoration to the original state is not possible in the case of large tissue losses. Thus, loss of functional skeletal muscle due to traumatic or degenerative conditions often results in deficits with poor treatment options [2]. Currently, tissue transplantation represents the only viable therapeutic alternative, which is associated with significant donor site morbidity [3]. *Ex vivo* engineered muscles, consisting of scaffolds containing differentiated muscle progenitor cells, may represent a viable alternative to replace or regenerate the damaged tissue. Although significant advances have been achieved in recent years, several practical challenges still remain. One major hurdle consists in procuring the appropriate amount of muscle progenitor cells [4]. Another important requirement is to establish the optimal conditions for cell proliferation, maturation, and assembly of the skeletal muscle fibers [5]. Finally, clinically relevant amounts of tissue will require means to provide for vascularization and innervation [6, 7]. In this chapter, we will focus on current approaches to efficiently organize and differentiate skeletal muscle fibers, which represents the primary step toward development of fully functional skeletal muscle tissue.

5.2 *In Vitro* Models of Adult Myogenesis

Adult skeletal muscle regeneration relies on a population of resident mononucleated stem cells known as the satellite cells. Quiescent satellite cells are located immediately under the basal lamina of myofibers and are readily activated by muscle damage. By undergoing asymmetric divisions, they contribute in the replenishment of the satellite cell compartment and formation of new, functional myofibers [8, 9]. Most of the current knowledge concerning skeletal muscle regeneration has been obtained through highly controllable and well-defined *in vitro* models, designed to investigate the role of microenvironmental cues in myogenic cell proliferation and differentiation. Although satellite cells isolated from healthy or diseased muscle tissue have been employed,

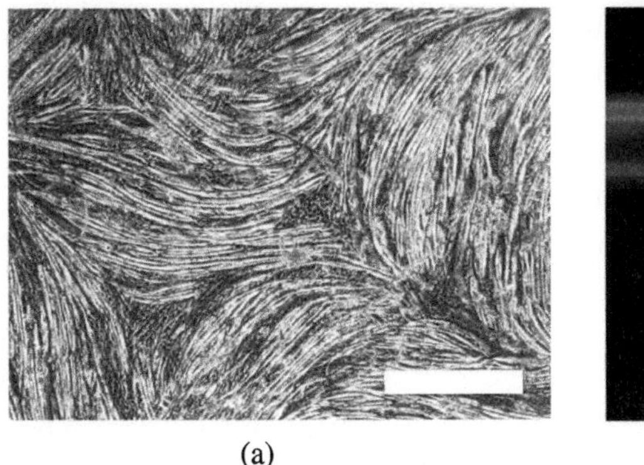

(a)

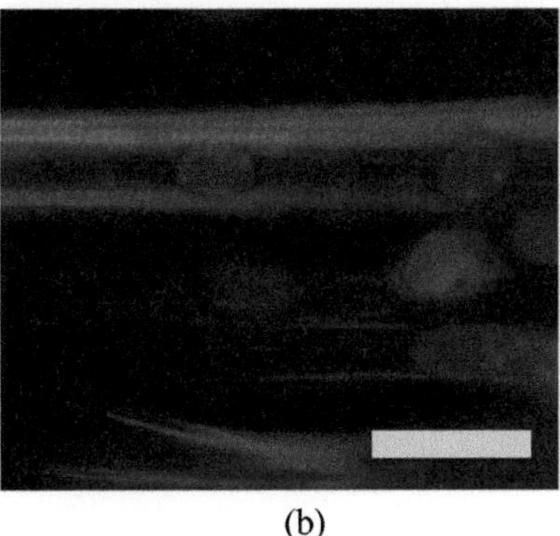

(b)

Figure 5.1 Skeletal muscle myotubes obtained from the C2C12 cell line. (a) Phase contrast microscopy image displaying the arbitrary orientation of the myotubes. Scale bar denotes 300 μm. (b) Fluorescence microscopy image showing myofibrils in mature myotubes. The actin filaments are stained by phalloidin (green) and nuclei counterstained by Hoechst (blue). Scale bar denotes 50 μm.

primary myogenic cultures possess various limitations, including cellular heterogeneity and low replicative capacity [10]. Therefore, the most widely used cell line for *in vitro* studies has been the immortalized mouse myoblast cell line C2C12. These cells display a large proliferation rate in high-serum conditions, and readily differentiate and fuse into myotubes upon reduction of serum mitogens (Figure 5.1a). Mature myofibers displaying actin-myosin cross-striations and contractile capacity are usually found after 5 to 7 days of induction (Figure 5.1b). Additionally, mesenchymal stem cells (MSCs) have been investigated for application in skeletal muscle tissue engineering. However, their ability to undergo terminal myogenic differentiation remains limited. Currently, diverse strategies are being employed to enhance their myogenic potential, including cyclic mechanical stimulation, coculture with other myogenic precursors, and hypoxic preconditioning [11–13]. While various types of skeletal muscle progenitors can be cultured and differentiated *in vitro*, without appropriate engineering strategies these cells only differentiate into poorly organized and nonfunctional arrays of myotubes.

5.3 Effect of Soluble and Bound Biochemical Cues

The process of muscle regeneration is highly regulated and the factors that contribute in this process are, among others, muscle stretch, trauma, neural stimulation, and soluble growth factors. The interplay of inflammatory cytokines, such as transforming growth factor-β (TGFβ), and soluble growth factors, such as fibroblast growth factor (FGF) and insulin-like growth factors (IGFs), defines muscle regeneration and repair outcome. Specialized components of the extracellular matrix (ECM), such as laminin and several forms of collagen, are also key mediators of regeneration process. The effects of soluble and bound factors regulating satellite cell activity are well studied and comprehensive reviews are readily available in the literature [8, 14, 15].

5.4 Regulation of Cell Fate by Passive Physical Cues

Apart from biochemical signals, several biophysical and structural cues are also part of the *in vivo* muscle microenvironment. In the following subsections, some of the approaches that have been exploited *in vitro* to control attachment, organization, and myogenic differentiation of myogenic precursors will be described.

5.4.1 Substrate Topography

Topographical features of the microenvironment influence migration and organization of the cells. This phenomenon has been termed contact guidance, in which physical shapes of the substrate induce alignment or directional growth of cells. Contact guidance has been investigated *in vitro* using culture substrates patterned with micrometer-sized structures such as fibers or grooves [16–19]. In particular, it has been found that microgrooved substrates can efficiently align skeletal myoblasts, supporting the formation of parallel arrays of muscle fibers [20–23]. As an example, Figure 5.2a displays a microgrooved substrate featuring tracks of 4 μm of width and 1 μm of depth. This kind of micropattern favors the parallel assembly of myoblasts, which upon fusion will result in the formation of a highly aligned layer of myotubes (Figure 5.2b). However, this approach has some limitations. For instance, alignment is limited to a single layer of cells, since cells that are not in direct contact with the substrate do not adopt the expected orientation [20]. Furthermore, topographical cues do not seem to provide any significant advantage for enhancement of the myogenic differentiation process in comparison to non-textured substrates [24, 25].

5.4.2 Substrate Stiffness

Aside from responding to structural cues, cells are able to sense the elasticity of their microenvironment. The dynamic tensional balance between the cell

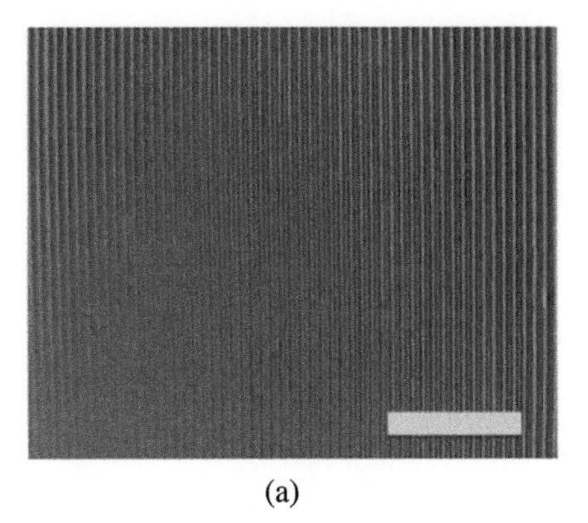

(a)

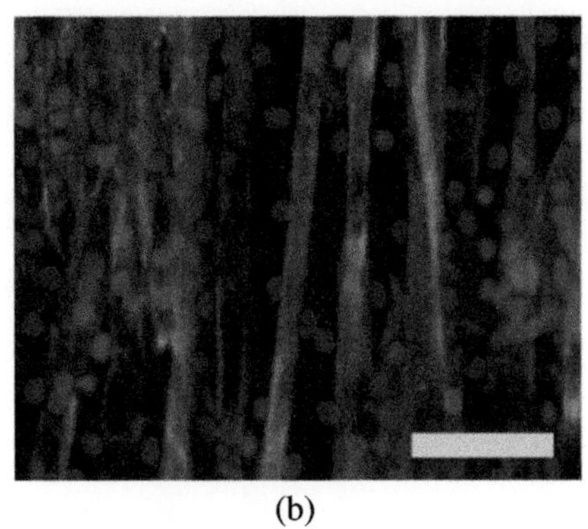

(b)

Figure 5.2 Organization of skeletal myotubes on micropatterned substrates. (a) Microgrooved silicone substrate used for cell patterning. Scale bar denotes 60 μm. (b) Fluorescence microscopy image showing C2C12 myotubes aligned on a micropatterned silicone substrate. The actin filaments are stained by phalloidin (orange) and nuclei counterstained by Hoechst (blue). Scale bar denotes 100 μm.

cytoskeleton and its environment determines the strain state of the cell at a given time. Cell migration is, for instance, activated by changes in the environmental stiffness that occur after diverse pathologic conditions. This property has been termed durotaxis or durokinesis, which has been exploited to efficiently control *in vitro* cell location and phenotype [26, 27]. In skeletal muscle, external stiffness is a key regulator of myogenic repair process. In particular, differentiation of skeletal myocytes is optimal in substrates matching the stiffness of muscles [28]. As shown by Monge et al., microstructured substrates for guiding cell adhesion and differentiation can be combined with thin films with tunable mechanical properties to devise optimal conditions for alignment and maturation of myotubes *in vitro* [29, 30].

5.5 Active Stimulation

The use of substrates with physical features that mimic the native environment has proven a valuable approach to control cell arrangement. However, in most of the cases, cues provided by the substrate are static in time and fail reproducing the dynamic characteristics of the *in vivo* environment. In the following subsections, some of the *in vitro* approaches used to provide dynamically changing cues will be described.

5.5.1 Electrical Stimulation

Electrical activity is part of the muscle cell niche *in vivo* and the effects of electric stimulation on skeletal muscles are well described. Application of electric fields of appropriate amplitude and frequency depolarize the membrane of muscle fibers and trigger a contractile response. Following this principle, electrical stimulation has been applied to enhance maturation of skeletal muscle cells *in vitro*. Diverse protocols have been proposed, leading to advanced myogenic differentiation, mainly in terms of increased synthesis of myosin heavy chain [31, 32]. However, the optimal stimulation parameters (duration, amplitude, frequency, etc.) remain largely unknown.

5.5.2 Mechanical Loading

Externally applied mechanical forces are also determinants of cell fate. Endothelial cells in blood vessels, for instance, are constantly exposed to a spectrum of hemodynamic forces caused by the pulsatile blood flow [33]. These forces include hydrostatic pressure, shear stress from the vessel wall, and cyclic strain, which all together determine the function and location of the cells. The link between the sensing of mechanical cues and the activation of cellular responses has been defined as mechanotransduction, which is fundamental for the maintenance of normal structure and function of various tissues [34]. In skeletal muscles, mechanotransduction is crucial, as mechanical forces control the development, maintenance, and repair of the tissue [35]. The sensitivity of muscle cells toward external mechanical loading has been demonstrated in diverse *in vitro* settings [36–38]. However, cultured muscle cells have been shown to respond differently to different types of mechanical stimulation paradigms. Cell responses depend, among others, on the rate, amplitude, and direction of the applied loading. Uniaxial tensile strain applied at a rate of few micrometers per minute has been shown to favor the alignment of muscle cells in the direction of the imposed strain [39]. On the other hand, the application of alternating phases of extension/relaxation, known as cyclic tensile strain (CTS), promotes a significant effect on mammalian myoblastic precursors, which respond by G-protein activation and increased protein synthesis [40]. In addition, cells are able to rearrange themselves on the culture substrate with the major axis aligned perpendicularly to the axis of the strain, following the principle of actin fiber reorganization (Figure 5.3). Remarkably, cell maturation is enhanced in response to uniaxial CTS, as evidenced by the presence of large numbers of myosin-positive myofibers [41].

Among the various mechanisms of mechanotransduction involved in this process, integrin mediated focal adhesions are believed to play a key role in transforming the externally applied forces in intracellular biochemical signals [42]. Integrin-mediated signaling initiates downstream activation of adaptor proteins such as the Rho family of GTPases and focal adhesion kinases (FAK). These signaling cascades are key activators of various transcription factors from the myogenic regulatory factor (MRF) family, involved in proliferation and differentiation of muscle precursors [43].

Although the application of mechanical stimulation in the field of skeletal tissue engineering is very promising, its full potential remains to be realized. Given that the outcomes of different stimulation paradigms have been contradicting, stimulation protocols have to be optimized in terms of frequency, amplitude, and duration to maximize the growth and maturation of skeletal muscle precursors. One of the fundamental questions is whether the effects observed in two-dimensional systems are also valid for the three-dimensional settings.

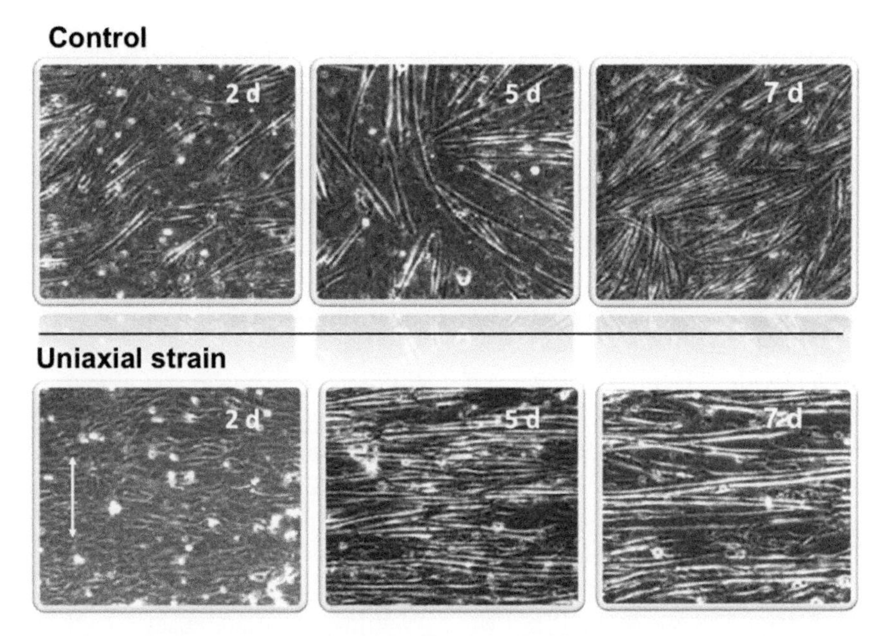

Figure 5.3 Muscle precursors after 2, 5 and 7 days of differentiation. The top row displays static C2C12 cultures. The bottom row shows the cells that were subjected to uniaxial CTS. The white arrow indicates the direction of the principal strain.

5.6 Summary and Perspectives

This review has attempted to provide a brief overview of *in vitro* studies useful to describe some fundamental aspects of the responses of skeletal muscle cells to environmental cues. These studies demonstrated that myoblasts differentiate optimally when the physical signals resemble the cues encountered by the cells in their natural environment. It is worth noting that the different physical signals cannot be completely separated from each other, since the extent of mechanical loading of the cells at a given point in time is determined by the balance between the external and the intrinsically generated forces. It is reasonable to assume that substrate stiffness, topography, applied strain, and even electrical stimulation trigger mechanotransduction mechanisms that activate downstream signaling cascades subserving identical functions. Thus, skeletal muscle engineering approaches aiming to develop highly organized cellular assemblies that mimic the natural skeletal muscle morphology necessarily need to take into account the contributions from all these physical cues.

Significant challenges remain along the way to establish tissue-engineered muscle as a viable therapeutic option. Future work should be focused on the development of novel skeletal muscle scaffolds providing appropriate mechanical and structural cues supporting muscle maturation. In addition, there is a need to investigate alternative cellular models that would overcome the limitations of using immortalized cell lines. Furthermore, efforts should be focused in investigating responses of skeletal muscle cells in more realistic, three-dimensional environments, since the principles established in two-dimensional cultures might not be applicable.

References

[1] R.R. Wolfe, The underappreciated role of muscle in health and disease, Am. J. Clin. Nutr. 84 (2006) 475–482.

[2] J. Huard, Y. Li, F.H. Fu, Muscle injuries and repair: current trends in research, J Bone Joint Surg Am. 84-A (2002) 822–832.

[3] B. Carlson, Skeletal muscle transplantation, in: R. Lanza, W. Chick (Eds.), Yearbook of Cell and Tissue Transplantation 1996–1997, Springer Netherlands, Dordrecht, 1996: pp. 61–67.

[4] G. Schaaf, F. Sage, M. Stok, E. Brusse, W. Pijnappel, A.J. Reuser, *Ex-vivo* Expansion of Muscle-Regenerative Cells for the Treatment of Muscle Disorders, J Stem Cell Res Ther. 2 (2012).

[5] M. Koning, M.C. Harmsen, M.J.A. van Luyn, P. Werker, Current opportunities and challenges in skeletal muscle tissue engineering, J Tissue Eng Regen Med. 3 (2009) 407–415.

[6] A.K. Saxena, G.H. Willital, J.P. Vacanti, Vascularized three-dimensional skeletal muscle tissue-engineering, Bio-Medical Materials and Engineering 11 (2001) 275–282.

[7] S. Levenberg, J. Rouwkema, M. Macdonald, et al., Engineering vascularized skeletal muscle tissue, Nat Biotechnol. 23 (2005) 879–884.

[8] E. Schultz, K.M. McCormick, Skeletal muscle satellite cells, Rev Physiol Biochem Pharmacol. 123 (1994) 213–257.

[9] S. Kuang, K. Kuroda, F. Le Grand, M.A. Rudnicki, Asymmetric self-renewal and commitment of satellite stem cells in muscle, Cell 129 (2007) 999–1010.

[10] S. Decary, V. Mouly, C.B. Hamida, A. Sautet, J.P. Barbet, G.S. Butler-Browne, Replicative Potential and Telomere Length in Human Skeletal Muscle: Implications for Satellite Cell-Mediated Gene Therapy, Human Gene Therapy 8 (1997) 1429–1438.

[11] J.-H. Lee, D.M. Kemp, Human adipose-derived stem cells display myogenic potential and perturbed function in hypoxic conditions, Biochem Biophys Res Commun. 341 (2006) 882–888.

[12] H. Egusa, M. Kobayashi, T. Matsumoto, J.-I. Sasaki, S. Uraguchi, H. Yatani, Application of cyclic strain for accelerated skeletal myogenic differentiation of mouse bone marrow-derived mesenchymal stromal cells with cell alignment, Tissue Eng Part A. 19 (2013) 770–782.

[13] J.I. Andersen, M. Juhl, T. Nielsen, et al., Uniaxial cyclic strain enhances adipose-derived stem cell fusion with skeletal myocytes, Biochemical and Biophysical Research Communications VL. 450 (2014) 1083–1088.

[14] S.B.P. Chargé, M.A. Rudnicki, Cellular and molecular regulation of muscle regeneration, Physiol. Rev. 84 (2004) 209–238.

[15] M. Karalaki, S. Fili, A. Philippou, M. Koutsilieris, Muscle regeneration: cellular and molecular events, In Vivo. 23 (2009) 779–796.

[16] P.T. Ohara, R.C. Buck, Contact guidance *in vitro*. A light, transmission, and scanning electron microscopic study, Exp Cell Res. 121 (1979) 235–249.

[17] D.M. Brunette, Fibroblasts on micromachined substrata orient hierarchically to grooves of different dimensions, Exp Cell Res. 164 (1986) 11–26.

[18] H.G. Craighead, C.D. James, A.M.P. Turner, Chemical and topographical patterning for directed cell attachment, Current Opinion in Solid State and Materials Science. 5 (2001) 177–184.
[19] J.C. Grew, J.L. Ricci, H. Alexander, Connective-tissue responses to defined biomaterial surfaces. II. Behavior of rat and mouse fibroblasts cultured on microgrooved substrates, J. Biomed. Mater. Res. 85 (2008) 326–335.
[20] T. Neumann, S.D. Hauschka, J.E. Sanders, Tissue Engineering of Skeletal Muscle Using Polymer Fiber Arrays, Tissue Engineering 9 (n.d.) 995–1003.
[21] T.M. Patz, A. Doraiswamy, R.J. Narayan, R. Modi, D.B. Chrisey, Two-dimensional differential adherence and alignment of C2C12 myoblasts, Materials Science and Engineering: B. 123 (2005) 242–247.
[22] J. Gingras, R.M. Rioux, D. Cuvelier, et al., Controlling the orientation and synaptic differentiation of myotubes with micropatterned substrates, Biophys. J. 97 (n.d.) 2771–2779.
[23] L. Altomare, N. Gadegaard, L. Visai, M.C. Tanzi, S. Farè, Biodegradable microgrooved polymeric surfaces obtained by photolithography for skeletal muscle cell orientation and myotube development, Acta Biomater. 6 (2010) 1948–1957.
[24] J.L. Charest, A.J. García, W.P. King, Myoblast alignment and differentiation on cell culture substrates with microscale topography and model chemistries, Biomaterials 28 (2007) 2202–2210.
[25] K. Shimizu, H. Fujita, E. Nagamori, Alignment of skeletal muscle myoblasts and myotubes using linear micropatterned surfaces ground with abrasives, Biotechnol. Bioeng. 103 (2009) 631–638.
[26] C.-M. Lo, H.-B. Wang, M. Dembo, Y.-L. Wang, Cell Movement Is Guided by the Rigidity of the Substrate, Biophys. J. 79 (2000) 144–152.
[27] D.S. Gray, J. Tien, C.S. Chen, Repositioning of cells by mechanotaxis on surfaces with micropatterned Young's modulus, J. Biomed. Mater. Res. 66 (2003) 605–614.
[28] A.J. Engler, Myotubes differentiate optimally on substrates with tissue-like stiffness: pathological implications for soft or stiff microenvironments, The Journal of Cell Biology 166 (2004) 877–887.
[29] C. Monge, K. Ren, R. Guillot, K. Berton, D. Peyrade, C. Picart, Engineering muscle tissues on microstructured polyelectrolyte multilayer films, Tissue Eng Part A. (2012) 120518104605000.

[30] C. Monge, N. Saha, T. Boudou, et al., Rigidity-Patterned Polyelectrolyte Films to Control Myoblast Cell Adhesion and Spatial Organization – Monge – 2013 – Advanced Functional Materials – Wiley Online Library, Adv. Funct. Mater. 23 (2013) 3432–3442.

[31] J. Stern-Straeter, A.D. Bach, L. Stangenberg, et al., Impact of electrical stimulation on three-dimensional myoblast cultures – A real-time RT-PCR study, Journal of Cellular and Molecular Medicine 9 (n.d.) 883–892.

[32] M.L.P. Langelaan, K.J.M. Boonen, K.Y. Rosaria-Chak, D.W.J. van der Schaft, M.J. Post, F.P.T. Baaijens, Advanced maturation by electrical stimulation: Differences in response between C2C12 and primary muscle progenitor cells, J Tissue Eng Regen Med 5 (2010) 529–539.

[33] M.A. Gimbrone, T. Nagel, J.N. Topper, Biomechanical activation: an emerging paradigm in endothelial adhesion biology, J. Clin. Invest 99 (1997) 1809–1813.

[34] K.S. Kolahi, M.R.K. Mofrad, Mechanotransduction: a major regulator of homeostasis and development, Wiley Interdisciplinary Reviews: Systems Biology and Medicine. 2 (2010) 625–639.

[35] G. Goldspink, Mechanical signals, IGF-I gene splicing, and muscle adaptation, Physiology (Bethesda) 20 (2005) 232–238.

[36] H.H. Vandenburgh, S. Hatfaludy, P. Karlisch, J. Shansky, Mechanically induced alterations in cultured skeletal muscle growth, J Biomech. 24 Suppl 1 (1991) 91–99.

[37] S. Abe, S. Rhee, O. Iwanuma, et al., Effect of mechanical stretching on expressions of muscle specific transcription factors myod, Myf-5, myogenin and MRF4 in proliferated myoblasts, Journal of Veterinary Medicine Series C: Anatomia Histologia Embryologia 38 (n.d.) 305–310.

[38] R.J. Segurola, B.E. Sumpio, I. Mills, Strain-induced dual alignment of L6 rat skeletal muscle cells, *In Vitro* Cell. Dev. Biol. – Animal 34 (1998) 609–612.

[39] H.H. Vandenburgh, P. Karlisch, Longitudinal growth of skeletal myotubes *in vitro* in a new horizontal mechanical cell stimulator, *In Vitro* Cell. Dev. Biol.-Animal 25 (n.d.) 607–616.

[40] C.B. Clark, N.L. McKnight, J.A. Frangos, Stretch activation of GTP-binding proteins in C2C12 myoblasts, Exp Cell Res. 292 (n.d.) 265–273.

[41] C.P. Pennisi, C.G. Olesen, M. de Zee, J. Rasmussen, V. Zachar, Uniaxial cyclic strain drives assembly and differentiation of skeletal myocytes, Tissue Eng Part A 17 (2011) 2543–2550.
[42] W.W. Ahmed, T. Wolfram, A.M. Goldyn, et al., Myoblast morphology and organization on biochemically micro-patterned hydrogel coatings under cyclic mechanical strain, Biomaterials 31 (2010) 250–258.
[43] S.J. Zhang, G.A. Truskey, W.E. Kraus, Effect of cyclic stretch on beta1D-integrin expression and activation of FAK and RhoA, AJP: Cell Physiology 292 (2007) C2057–C2069.

6

Effect of Bioactive Growth Surfaces on Human Mesenchymal Stem Cells: A Pilot Biomarker Study to Assess Growth and Differentiation

Liliana Craciun[1], Pranela Rameshwar[2], Steven J. Greco[2], Ted Deisenroth[1] and Carmen Hendricks-Guy[1]

[1]BASF Corporation, 500 White Plains Road, Tarrytown, NY 10591, USA
[2]New Jersey Medical School, Department of Medicine – Division of Hematology/Oncology, Rutgers School of Biomedical Health Science, Newark, NJ 07103, USA

Abstract

Mesenchymal stem cells (MSCs) are multipotent adult stem cells with the ability to differentiate into multiple cell lineages, and possess significant potential towards application in cell therapy and tissue engineering. A significant barrier to the effective implementation of human MSC (hMSC) therapies is the limited access to large quantities of viable, homogeneous cell populations produced in reproducible, consistent cultures. hMSCs are adherent cells whose morphology, as well as proliferation and differentiation potential, depend on the characteristics of the tissue culture surface they grow on. In the present study, we screened a panel of plastic surfaces possessing various physical characteristics for their ability to expand and maintain hMSCs in an undifferentiated state. The purpose of these studies was to identify materials which outperform conventional tissue culture polystyrene (TCPS). The plastic surfaces that were investigated were created by injection molding and were used either "as is" or after plasma treatment under an oxidative or reductive atmosphere. After 5 days culture, the effect of the growth surfaces on stem cell maintenance and differentiation was quantified by the expression of stem cell markers (OCT-4; NANOG; NOTCH1; PH-4; p21).

Several surfaces exhibited an increase in the stem cell specific- and/or a decrease in the differentiation-specific genes, indicating a "positive result" compared to the TCPS reference standard. Of the many surface physical characteristics, the roughness and fibrous structure of a glass fiber-reinforced poly(styrene-acrylonitrile) polymeric surface had the most prevalent effect on facilitating hMSC expansion with preservation of stem cell function. These findings are not only significant in defining ideal conditions for hMSC growth in culture, but have broader implications for tissue engineering.

Keywords: Human Mesenchymal Stem Cells, Bioactive surface, Cell growth, Cell differentiation.

6.1 Introduction

Cell therapies show great promise for repairing or regenerating damaged cells, tissues and organs. Cells can carry out functions that cannot be performed by small-molecule drugs. They are adaptable and can sense their surroundings and vary their responses to better suit physiologic conditions. In particular, stem cells have both the long term capacity to replicate themselves, thereby maintaining a continuous cell supply, as well as the ability to differentiate into specialized cell types. By leveraging the capacity for self-renewal and regeneration, stem cell therapies, also referred to as regenerative or reparative medicine, offer hope for solving critical, unmet needs for a multitude of diseases and disorders, many of which are currently untreatable.

The therapeutic use of stem cells has been ongoing for several decades in the form of bone marrow (BM) transplants to treat various hematological disorders and immune-related diseases, and is a very active area of investigation [1–6]. In mammals there are two broad types of stem cells, embryonic stem cells (ESC) and non-embryonic or "adult" stem cells. The primary role of adult stem cells in living organisms is to maintain and repair the tissue in which they are found. Mesenchymal stem cells (MSC) are multi-potent adult stem cells available in bone marrow and adipose tissue [7–9]. They can differentiate into cell types such as adipocytes, osteoblasts, chondrocytes, cardiomyocytes, or neuronal cells, supporting the formation of blood and fibrous connective tissue [10–13]. MSC represent an ideal source for cellular replacement therapies because of their relative ease of isolation, high *in vitro* expansion rate, and demonstrated multipotency. In addition, MSCs suppress immune system rejection in individuals receiving them, increasing the likelihood of treatment success [14–22]. These properties circumvent host immune response issues allowing allogeneic cell sources which could be immediately available as an off-the-shelf therapy. Another advantage of MSCs is a lower probability of tumor formation compared to ESCs [4, 23, 24].

There is great interest in applications of MSCs in cell therapy and tissue engineering. MSCs are being tested for a variety of disorders with more than 360 active clinical trials in the US alone, where MSCs are evaluated in diseases including graft-versus-host disease, Crohn's disease, myocardial infarction, colitis, diabetes, cartilage defects, bone cysts, limb ischemia, Parkinson, arthritis, anemia, stroke, nephropathy, and many others. The first MSC drug therapy was approved by Canada in 2012 to treat children in graft vs. host disease, a nearly fatal complication arising during bone marrow transplantation.

However, using MSCs for medical treatments still poses problems that affect their clinical usefulness. Major challenges include the need to ensure safety, efficacy, consistent performance, and the ability to produce large quantities of homogeneous cell populations needed for clinical applications and treatment. MSCs must be amplified in culture by repeated passaging to create enough viable cells, however, prolonged expansion could potentially reduce the ability of the cells to differentiate. In cell-based therapies, cells are removed from the patient or a healthy donor and cultured in the laboratory where they are expanded before being infused into the patient. Cells expanded outside of their natural environment in the human body can become ineffective or produce adverse effects. MSCs are adherent cells whose cell morphology, proliferation and differentiation potential are affected by the surface they are grown on. To this end, the purpose of the present study was to investigate synthetic materials suitable for either supporting the growth and maintenance of MSCs in undifferentiated state or facilitating stem cell differentiation into specialized tissues. Herein, we screened a panel of plastic surfaces with different physical characteristics in order to identify materials which outperform conventional tissue culture polystyrene (TCPS). The plastic surfaces were created by injection molding and were used either "as is" or after plasma treatment under an oxidative or reductive atmosphere. Several surfaces exhibited a statistically significant increase in the stem cell specific-genes and/or a decrease in the differentiation-specific genes, indicating a "positive result" compared to the TCPS reference standard. Of the many surface physical characteristics the roughness and fibrous structure of a glass fiber-reinforced poly(styrene-acrylonitrile) polymeric surface had the most prevalent effect on facilitating hMSC expansion with preservation of stem cell function.

6.2 Materials and Methods

6.2.1 Materials

The polymer materials were sourced from BASF (Ultraform N 2320 003 Q600, Ultrason S 2010, UltraPET PCS Clear, Ultramid B27 E, Terluran GP-22, Terlux 2802, Terlux 2812, Styrolux 656 C, Styrolux 3G 46, Luran 378 P, Luran 378 PG-7) and Topas Advanced Polymers (Topas 6013S-04). Polymer chemistries and abbreviations are shown in Table 6.1.

Polymer processing was done by injection molding into 1 mm thick plaques using 170-tons Van Dorn (POM, PSU, PET, PA6) and Boy 50 M (ABD, MABS, SBC, SAN, COC) horizontal injection molding machines (BASF Engineering Plastics; Budd Lake and Tarrytown Laboratories). The polymer plaques had either a matte (rough) or a glossy (smooth) finish. They were cut into round coverslips of 35 mm diameter that fit into 6-well plates for cell culture. In addition the coverslips were chemically modified by plasma treatment. All samples were carefully cleaned from contaminants by washing with organic solvents and water in an ultrasonic bath. During plasma treatment the coverslips were exposed to atmospheric chemical plasmas of different compositions at 25 WD (500 W@11 fpm) using Enercon Tangential Plasma3 technology with a 1 mm electrodes air gap (Enercon Industries, Menomonee Falls, WI). The treatment conditions were: (a) 90% (90% N_2 + 10% H_2) + 10% O_2; (b) 90% N_2 + 10% NH_3; and (c) 80% helium + 20% O_2. It is expected that the reductive conditions (a & b) would increase the nitrogen content at

Table 6.1 Materials utilized in study

Abbreviation	Chemical Name	Trade Name
POM	Poly(oxymethylene)	Ultraform N 2320 003 Q600
PSU	Polysulfone	Ultrason S 2010
PET	Poly(ethylene terephthalate)	UltraPet PCS Clear
PA6	Polyamide 6; poly(ε-caprolactam); nylon 6	Ultramid B27 E
ABS	Poly(acrylonitrile-1,3-butadiene-styrene)	Terluran GP-22
MABS	Poly(methyl methacrylate-acrylonitrile-1,3-butadiene-styrene)	Terlux 2802; Terlux 2812
SBC	Polystyrene-*block*-poly(1,3-butadiene)	Styrolux 656 C; Styrolux 3G 46
SAN	Poly(styrene-acrylonitrile)	Luran 378 P; Luran 378 PG-7
COC	Cyclo olefinic co-polymer; polyethylene-*block*-polynorbornene	Topas 6013S-04

the surface, whereas the oxidative plasma (c) would oxidize the surface and potentially make it rougher.

The characterization of the polymeric surfaces before and after plasma treatment was done by contact angle with water, atomic force microscopy (AFM), and X-ray photoelectron spectroscopy (XPS). AFM analysis was performed with a Dimension V scanning probe microscope from Vecco, used in tapping mode with a Bruker TESP tip. Contact angle analysis was performed on an OCA 20 goniometer from Future Digital Scientific Corporation. XPS analysis was done with a K-Alpha X-ray Photoelectron Spectrometer from Thermal Fisher. Samples were mounted on a standard sample holder using clips to hold the materials in place. An X-ray spot size of 400 μm was analyzed using an ion gun current of 3000 eV with a sputter rate of 0.23 nm/sec.

The bulk elemental analysis for carbon, hydrogen and nitrogen was done by combustion followed by microchemical techniques. The % oxygen is calculated as the difference to 100%. The values reported are averages of duplicate runs. CHN Analysis is a form of Elemental Analysis concerned with determination of only Carbon (C), Hydrogen (H) and Nitrogen (N) in a sample. The most popular technology behind the CHN analysis is combustion train analysis where the sample is first fully combusted and then the products of its combustion are analyzed. The full combustion is usually achieved by providing abundant oxygen supply during the combustion process. The analyzed products, Carbon, Hydrogen and Nitrogen, oxidize and form carbon dioxide (CO_2), water, and nitric oxide (NO), respectively. These product compounds are carefully measured. The gases in individual traps for CO_2 and water are measured for thermal conductivity before and after combustion. The concentrations are used to determine the elemental composition, or *empirical formula*, of the analyzed sample.

6.2.2 Cell Culture Reagents

Dulbecco's modified Eagle's medium (DMEM) with high glucose, trypsin-EDTA and α-MEM was purchased from Gibco (Grand Island, NY), and fetal calf serum (FCS) from Hyclone Laboratories (Logan, UT).

6.2.3 Culture of Human MSCs

MSCs were cultured from BM aspirates as described [14]. The use of human BM aspirates followed a protocol approved by the Institutional Review

Board of The University of Medicine and Dentistry of New Jersey-Newark campus. Unfractionated BM aspirates (2 ml) were diluted in 12 ml of DMEM containing 10% FCS (D10 media) and then transferred to vacuum-gas plasma treated, tissue culture Falcon 3003 petri dishes. Plates were incubated, and at day 3, mononuclear cells were isolated by Ficoll Hypaque density gradient and then replaced in the culture plates. Fifty percent of media was replaced with fresh D10 media at weekly intervals until the adherent cells were approximately 80% confluent. After four cell passages, the adherent cells were asymmetric, $CD14^-$, $CD29^+$, $CD44^+$, $CD34^-$, $CD45^-$, $SH2^+$, prolyl-4-hydroxylase$^-$ [14].

6.2.4 qPCR for Stem Cell Markers

Test surfaces were divided into three experimental groups which were analyzed independently. Tissue-culture treated polystyrene (BD Falcon) and BD PureCoat amine multi-well plates from BD Biosciences were used as references. Sterilization was done by UV-exposure before cell seeding. MSC were cultured from bone marrow aspirates of healthy donors, aged 20–35. After 5 days of culture, cells were harvested from each surface using enzymatic detachment and pelleted by centrifugation. Total RNA (2 μg) was reverse transcribed, and 200 ng of cDNA was used in quantitative PCR (qPCR) with the Platinum SYBR Green qPCR SuperMix-UDG Kit (Invitrogen, Carlsbad, CA). qPCRs were normalized by amplifying the same sample of cDNA with primers specific for β-actin. qPCRs were performed with a 7500 Real Time PCR System (Applied Biosystems, Foster City, CA). The cycling profile for real-time PCR (40 cycles) was as follows: 94°C for 15 seconds and 60°C for 45 seconds. Gene expression analysis was performed using the 7500 System SDS software (Applied Biosystems). Normalizations were performed with β-actin, and values were arbitrarily assigned a value of 1. Primer sequences and additional information are as follows: **OCT4** (NM_002701, +789/+1136) Forward: gtt cag cca aaa gac cat, Reverse: cgt tgt gca tag cca ctg; **SOX2** (NM_003106, +734/+1113) Forward: aag gag cac ccg gat tat, Reverse: tgc gag tag gac atg ctg; **NANOG** (NM_024865, +686/1016) Forward: act ggc cga aga ata gca, Reverse: aaa gca gcc tcc aag tca; **PH-4** (NM_177939, +810/+1166) Forward: aag agt gtc ggc tca tca, Reverse: cac cag ctc act gga ctc; **p21** (NM_000389, +904/+1324) Forward: gcc agc tac ttc ctc ctc, Reverse: aag agg gaa aag gct caa; **β-Actin** (NM_001101, +842/+1037) Forward: tgc cct gag gca ctc ttc, Reverse: gtg cca ggg cag tga tct.

6.3 Results

Polymeric coverslips were made by injection molding with either a matte (M1 or M2) or a glossy (G) finish and treated with high density discharge atmospheric plasma using mixtures of air (nitrogen, oxygen) with an inert gas (helium) or other reactive gases (hydrogen, ammonia) for surface functionalization (Figure 6.1). The coverslips were run in an Enercon Plasma3 system above the lower electrode connected to the ground (Table 6.2). The air gap between the electrodes is occupied by the glowing reactive gases. Three different gaseous mixtures were used: a) 90% (90% N_2 + 10% H_2) + 10% O_2; (b) 90% N_2 + 10% NH_3; and (c) 80% helium + 20% O_2. The resulting plasma interacts with the surface inducing chemical and physical modifications. The effect of plasma on a given material is determined by the chemistry of the reactions between the surface and the reactive species present in the plasma. Each gas produces a unique plasma composition and results in different surface properties. Oxidative conditions (a & c) are known to result in formation of hydroxyl, aldehyde, ketone, and carboxyl groups at the surface [25]. Plasma treatment with ammonia gas (b) gives mainly amine-containing surfaces [26, 27]. The chemistry changes at the surface of the plasma treated coverslips were analyzed with XPS. The elemental analysis XPS data is shown in Tables 6.3 and 6.4.

The treated surfaces were characterized by contact angle with water before and immediately after plasma treatment. The contact angles for all surfaces decreased significantly after plasma treatment relative to the untreated surfaces. The values are recorded in Table 6.2. In addition, the surface aging effects upon storage were evaluated by contact angle measurements two months after the plasma treatment. All surfaces had their contact angle increase to almost the initial value before treatment (Figure 6.2).

The matte and glossy surfaces had significantly different roughness. AFM was used to determine the mean surface roughness (RMS). The plasma treatments did not change the surface topography or roughness except for the POM and PA6 materials. The RMS values are also recorded in Table 6.2.

Overall, after injection molding and plasma treatment, the nine commercial polymers created 80 distinct surfaces, of which 60 unique combinations were tested for MSC growth (15 distinct polymers with glossy and mate surface finishes, and 3 different plasma treatments). BD TCPS Falcon and BD PureCoat Amine plates served as reference standards. The polymeric surfaces were first screened for their ability to maintain stem cell gene expression through the use of the stem cell markers: OCT-4, NOTCH1 and NANOG (Figure 6.3).

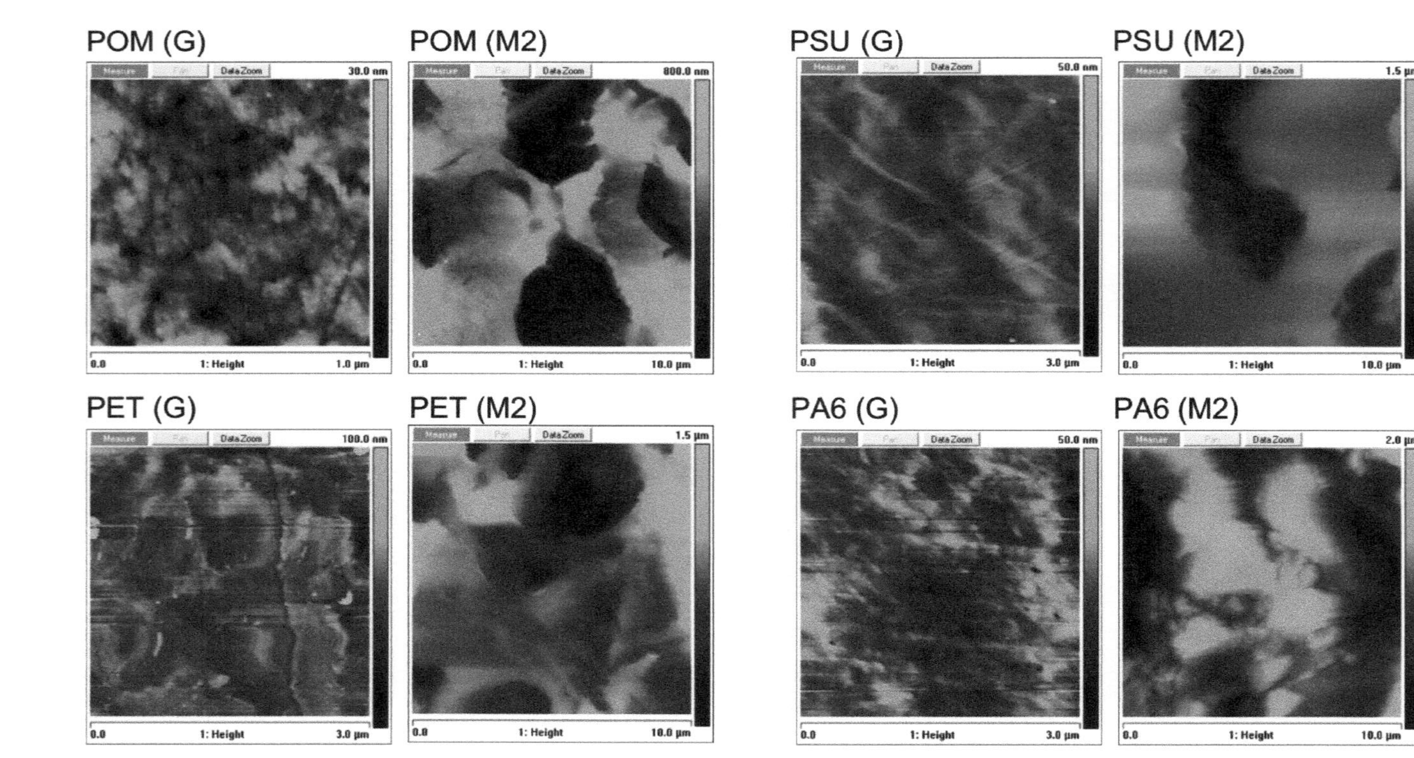

Figure 6.1 AFM phase images of plastic coverslips. AFM analysis was performed with a Dimension V scanning probe microscope from Vecco, used in tapping mode with a Bruker TESP vibrating cantilever tip. Sample height data is obtained from the changes in Z-axis displacement. The phase difference between the measured signal and the drive signal, caused by interactions between probe and material, gives the phase image, indicating regions of different composition and/or phase in the material.

Table 6.2 Characterization of plastic coverslips

No.	Surface	Modulus/Thermal	Surface Roughness[a]	Contact Angle[b] Initial	at 2 Months
1.	**POM** (G)	2.7 GPa; no T_g; T_m 168°C	NT: RMS = 3 nm	NT:82	76
				a:43	80
				b:48	68
				c:42	68
2.	**POM** (M1)	2.7 GPa; no T_g; T_m 168°C	NT: RMS = 171 nm	NT:74	75
				a:48	73
				b:56	72
			RMS = 262 nm	c:54	74
3.	**POM** (M2)	2.7 GPa; no T_g; T_m 168°C	NT: RMS = 244 nm	NT:70	70
				a:47	68
				b:50	74
				c:47	69
4.	**PSU** (G)	2.6 GPa; T_g 190°C	NT: RMS = 5 nm	NT:79	80
				a:30	56
				b:46	70
				c:34	84
5.	**PSU** (M1)	2.6 GPa; T_g 190°C	NT: RMS = 70 nm	NT:66	79
				a:39	68
				b:47	64
				c:47	70
6.	**PSU** (M2)	2.6 GPa; T_g 190°C	NT: RMS = 167 nm	NT:72	87
				a:34	75
				b:39	68
				c:49	73
7.	**PET** (G)	3.5 GPa; T_g 82°C	NT: RMS = 7 nm	NT:87	72
				a:47	60
				b:38	68
				c:37	73
8.	**PET** (M1)	3.5 GPa; T_g 82°C	NT: RMS = 194 nm	NT:80	81
				a:34	63
				b:45	83
			RMS = 177 nm	c:47	57
9.	**PET** (M2)	3.5 GPa; T_g 82°C	NT: RMS = 211 nm	NT:78	76
				a:44	64
				b:38	63
				c:40	68
10.	**PA6** (G)	1 GPa; no T_g; T_m 221°C	NT: RMS = 5 nm	NT:66	66
				a:30	72
				b:39	71
				c:38	61
11.	**PA6** (M1)	1 GPa; no T_g; T_m 221°C	NT: RMS = 339 nm	NT:80	79
				a:38	80
			RMS = 604 nm	b:49	76
				c:51	72
12	**PA6** (M2)	1 GPa; no T_g; T_m 221°C	NT: RMS = 434 nm	NT:78	77
				a:44	66
				b:48	71
				c:50	80

(*Continued*)

Table 6.2 Continued

No.	Surface	Modulus/Thermal	Surface Roughness[a]	Contact Angle[b] Initial	at 2 Months
13.	**ABS** (G)	2.3 GPa; T_g 109°C	NT: RMS = 56 nm	NT:74	84
				a:48	71
				b:50	60
				c:44	70
14.	**MABS 2802** (G)	2 GPa; 109°C	NT: RMS = 17 nm	NT:68	73
				a:50	68
				b:48	71
				c:52	71
15.	**MABS 2812** (G)	1.9 GPa; T_g 104°C	NT: RMS = 18 nm	NT:83	71
				a:48	73
				b:44	66
				c:55	71
16.	**SBC 656 C** (G)	1.8 GPa; T_g 105°C	NT: RMS = 5 nm	NT:84	85
				a:34	78
				b:42	81
			RMS = 3nm	c:29	75
17.	**SBC 3G 46** (G)	1.64 GPa; no T_g	NT: RMS = 8 nm	NT:80	94
				a:38	81
				b:41	79
				c:47	82
18.	**SAN 378 P** (G)	3.8 GPa; T_g 110°C	NT: RMS = 7 nm	NT:83	74
				a:35	68
				b:42	55
				c:34	62
19.	**SAN 378 PG-7** (G)	12 GPa; T_g 113°C	NT: RMS = 235 nm	NT:85	82
				a:33	58
				b:39	48
			RMS = 232 nm	c:35	61
20.	**COC** (G)	COC	NT: RMS = 11 nm	NT:90	91
		2.9 GPa; T_g 139°C		a:43	79
				b:46	64
				c:34	79
21.	**TCPS**	3–3.5 GPa T_g 100°C	RMS = 4 nm		
22.	**BD PureCoat Amine**	coated PS	RMS = 4 nm	NT:77	

[a]nd = not yet determined. NT = not treated or "as is". RMS = surface roughness. [b]Contact angle of water. Treatment conditions: (a) 90% (90% N_2 + 10% H_2) + 10% O_2; (b) 90% N_2 + 10% NH_3; and (c) 80% helium + 20% O_2.

After 5 days of culture, MSCs displayed an increase in the stem cell-specific markers compared to reference for a number of the initial surfaces tested. This result led us to investigate whether similar effects were observed with the remainder of the surfaces created. Again, we assessed stem cell-specific markers (Figure 6.4a) as well as markers of differentiation (p21, PH4) (Figure 6.4b). For surfaces that were able to be imaged by light microscopy,

Table 6.3 XPS and bulk elemental analysis of plastic coverslips

Sample/Element	C [wt %]	H [wt %]	N [wt %]	O [wt %]	Ratio O/C
MABS 2802					
Bulk elemental analysis	78.4	8.7	1.9	11.0	0.14
G-a	79.1		2.0	17.0	0.21
G-b	75.3		4.0	18.4	0.24
G-c	79.8		2.4	17.1	0.21
ABS GP-22					
Bulk elemental analysis	86.6	8.2	5.3	0	0
G-NT	91.3		4.0	4.7	0.05
G-b	81.6		7.0	10.5	0.13
G-c	84.7		4.7	10.3	0.13
POM					
Calculated elemental analysis	40.0	6.7	0	53.3	1.33
M1-a	52.6		0.7	42.1	0.80
M1-c	51.0		1.2	44.7	0.88
SAN 378 P					
Bulk elemental analysis	84.4	7.2	8.5	0	0
G-a	75.1		5.2	15.0	0.20
G-b	78.4		7.3	12.1	0.15
G-c	81.2		6.1	11.1	0.14
SBC 656 C					
G-NT	97.4		0.9	1.6	0.02
G-a	80.3		1.1	15.2	0.19
G-b	77.5		3.5	14.2	0.18
G-c	68.3		0.8	22.7	0.33
BD PureCoat Amine	79.6		6.5	12.7	0.16

Table 6.4 XPS data analysis of SBC 656 C coverslips

Sample/Element ID[a,b]	SBC 656 C	SBC 656 C	SBC 656 C	SBC 656 C
Element/ID[1]	G-NT	G-a	G-b	G-c
C CH_x	89.2	53.0	68.3	57.1
C CO	4.4	16.5	5.2	2.6
C CO_2 + CO_3	1.0	2.4	3.3	7.6
C aromatic, C(halide)	2.6	1.0	0.7	1.0
N nitride, NH_3	1.0	1.0	3.5	0.8
N NO, NH_4	nd	0.6	nd	nd
O	1.8	20.5	14.2	22.7

[a]Chemical state identifications are based on consistencies between reference and measured binding energies and are not absolute. [b] nd = not detected.

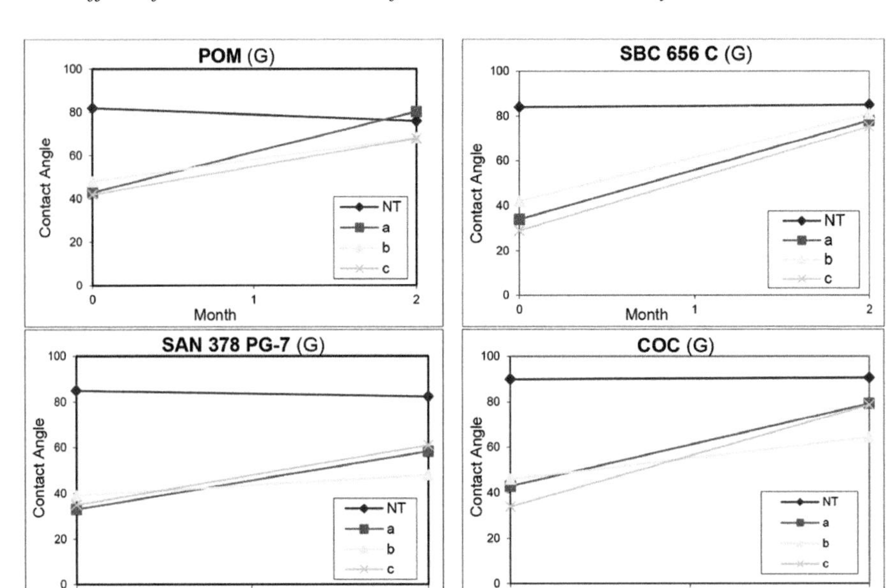

Figure 6.2 Water contact angle of POM, SBC 656 C, SAN 378 PG-7, and COC coverslips immediately after (month = 0) and two months (month = 2) after plasma treatment. The analysis was performed on an OCA 20 goniometer from Future Digital Scientific Corporation, by the sessile drop method. The optical system captures the profile of the water droplet on the surface. The angle between the water/solid interface is the contact angle. A low contact angle with water means that the surface is hydrophilic. A surface with a high water contact angle, usually larger than 90°, is considered hydrophobic.

cell morphologies were additionally examined (Figure 6.5). From the 60 different surfaces evaluated, 12 performed significantly better than the polystyrene reference surface in up-regulating stem-cell specific and down-regulating differentiation-specific genes. Future studies are necessary to validate whether these test surfaces can also maintain long-term stem cell expansion, to discern any global changes in stem cell gene expression through microarray analyses and to confirm that functionality is maintained through examining lineage-specific differentiation.

6.4 Discussion

The scope of this work was to evaluate the ability of standard polymeric materials to serve as tissue-culture surfaces for expansion and maintenance of stem cell phenotype in hMSCs. Plasma-treated polystyrene (TCPS) is

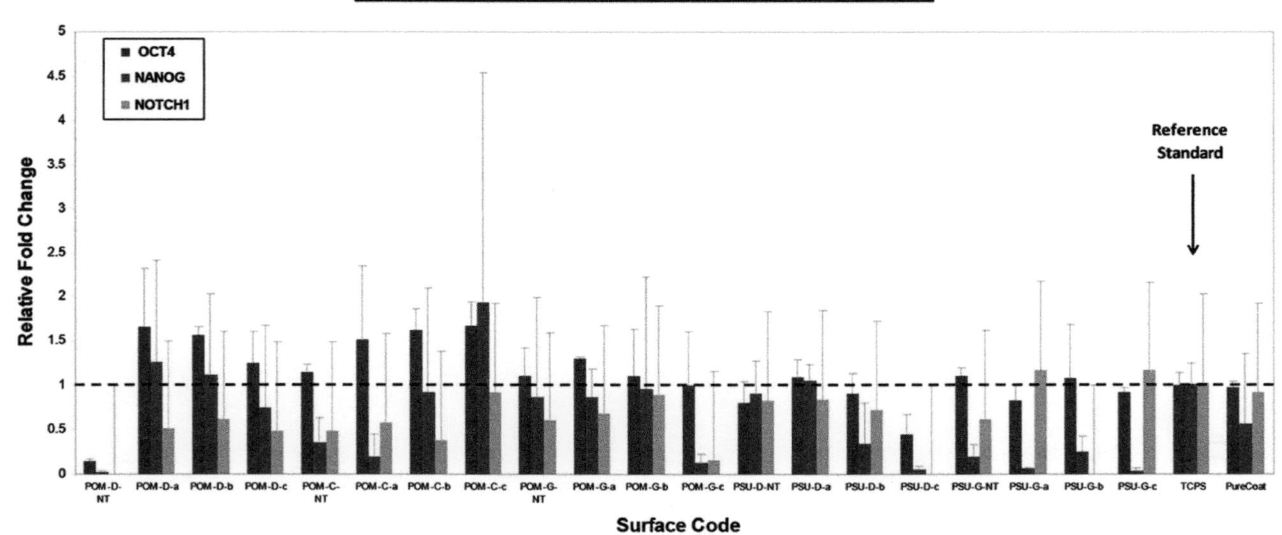

Figure 6.3 Effect of plastic surfaces on relative pro-stem cell gene expression in MSC cultures. After 5 days of culture, cells were harvested and RNA analyzed by qPCR for the stem cell genes, OCT4, NOTCH1 and NANOG. Normalizations were performed with β-actin, and values were arbitrarily assigned a value of 1 relative to BD TCPS Falcon. BD PureCoat Amine plates also served as a reference material.

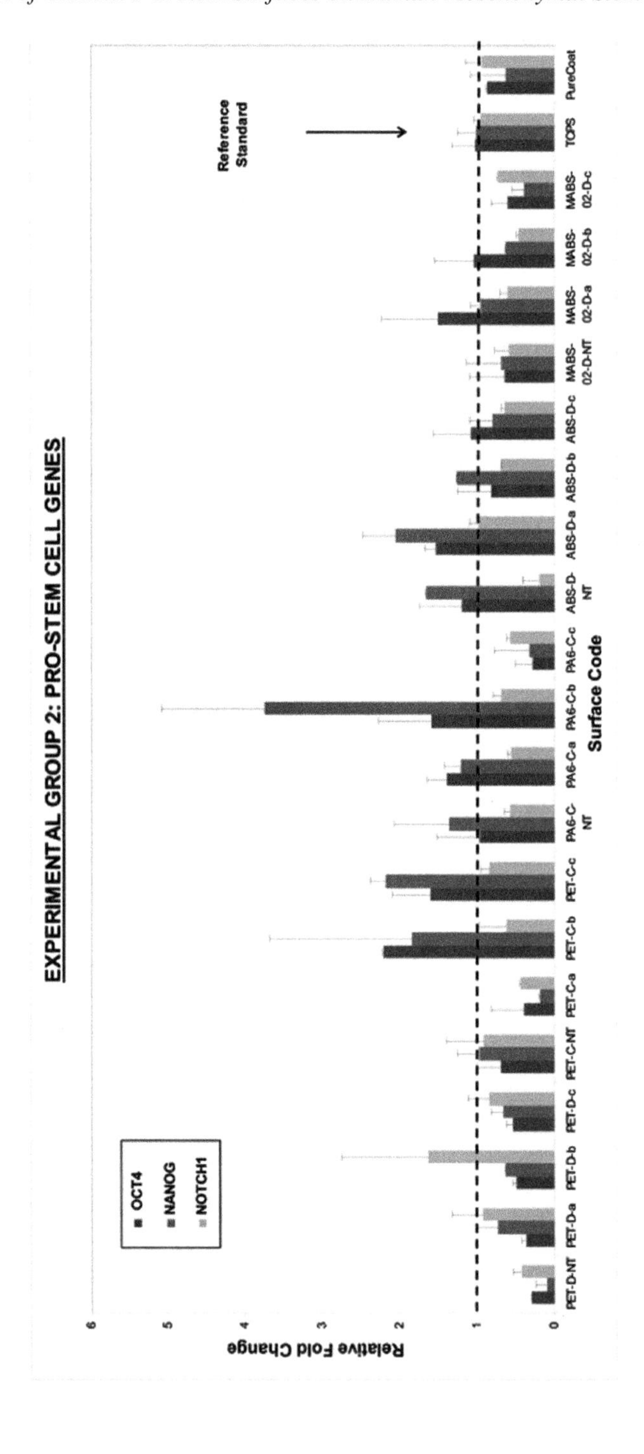
EXPERIMENTAL GROUP 2: PRO-STEM CELL GENES
OCT4
NANOG
NOTCH1
Reference Standard
Relative Fold Change
0
1
2
3
4
5
6
PET-D-NT
PET-D-a
PET-D-b
PET-D-c
PET-C-NT
PET-C-a
PET-C-b
PET-C-c
PA6-C-NT
PA6-C-a
PA6-C-b
PA6-C-c
ABS-D-NT
ABS-D-a
ABS-D-b
ABS-D-c
MABS-02-D-NT
MABS-02-D-a
MABS-02-D-b
MABS-02-D-c
TOPS
PureCoat
Surface Code

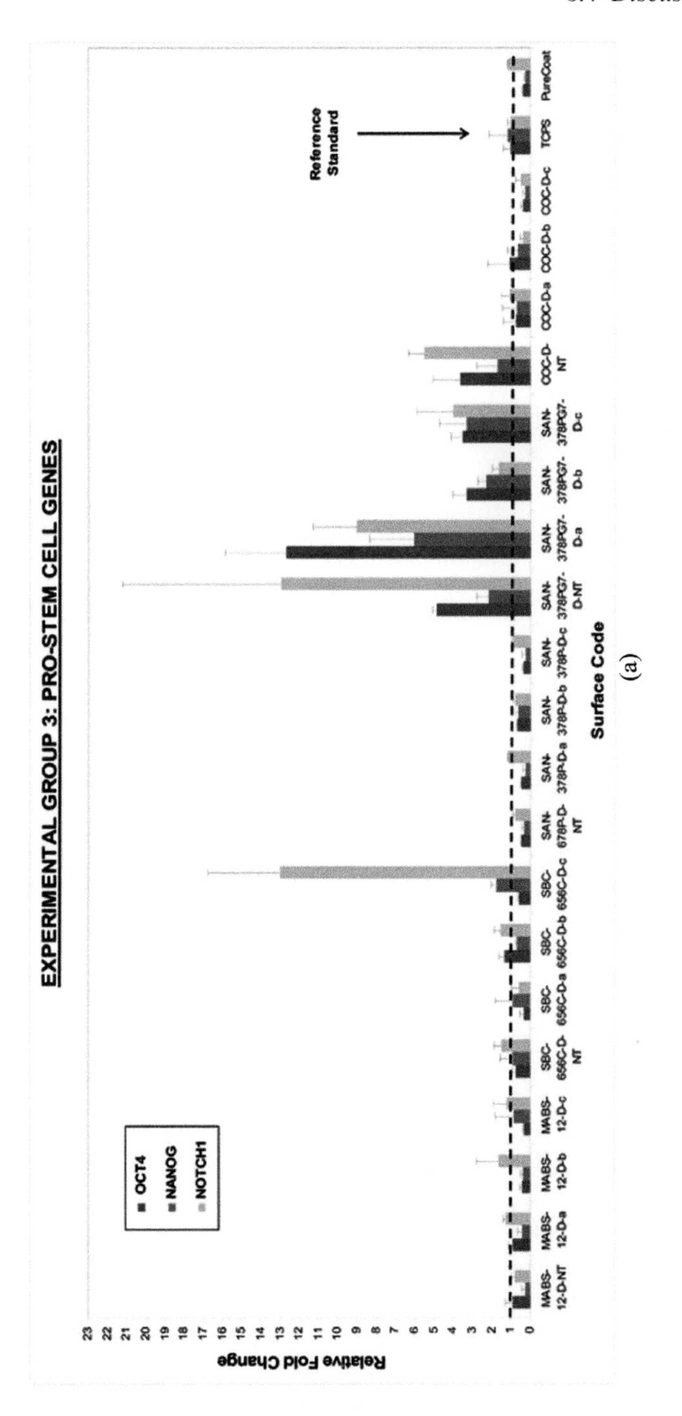
EXPERIMENTAL GROUP 3: PRO-STEM CELL GENES
OCT4
NANOG
NOTCH1
Relative Fold Change
0
1
2
3
4
5
6
7
8
9
10
11
12
13
14
15
16
17
18
19
20
21
22
23
Reference Standard
MABS-12-D-NT
MABS-12-D-a
MABS-12-D-b
MABS-12-D-c
SBC-656C-D-NT
SBC-656C-D-a
SBC-656C-D-b
SBC-656C-D-c
SAN-678P-D-NT
SAN-378P-D-a
SAN-378P-D-b
SAN-378P-D-c
SAN-378PG7-D-NT
SAN-378PG7-D-a
SAN-378PG7-D-b
SAN-378PG7-D-c
COC-D-NT
COC-D-a
COC-D-b
COC-D-c
TOPS
PureCoat
Surface Code

(a)

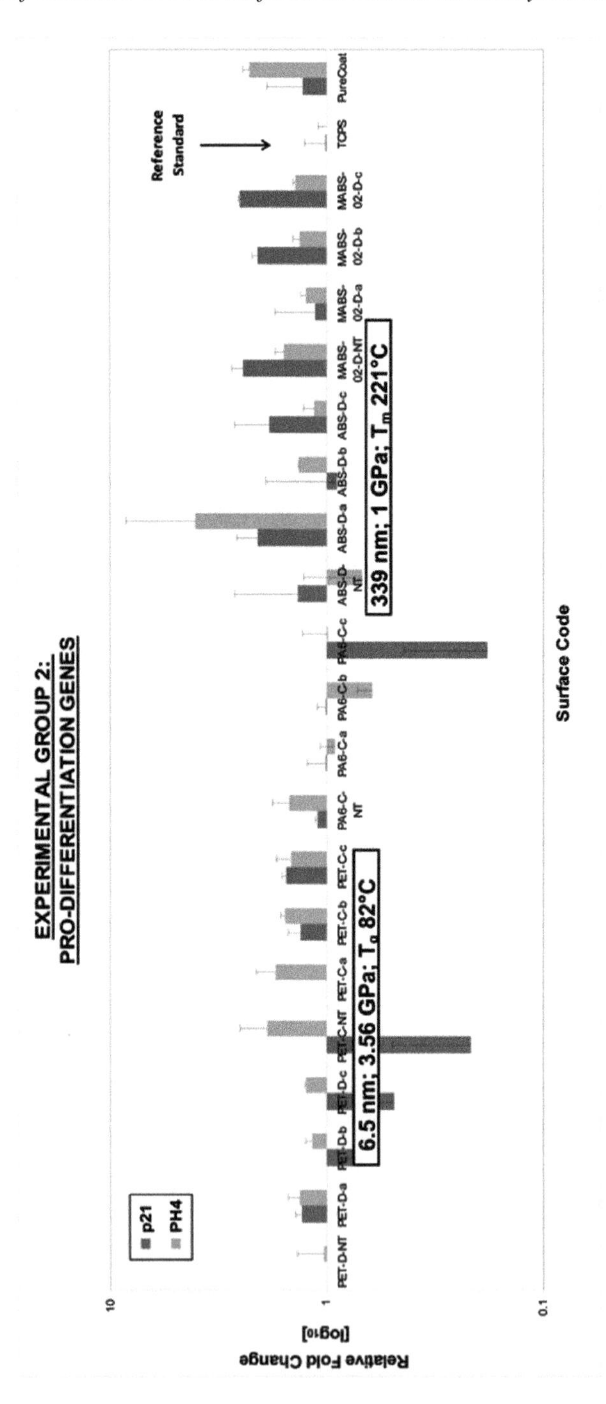
EXPERIMENTAL GROUP 2:
PRO-DIFFERENTIATION GENES
p21
PH4
Reference Standard
6.5 nm; 3.56 GPa; T_g 82°C
339 nm; 1 GPa; T_m 221°C
Relative Fold Change [log10]
10
1
0.1
PET-D-NT
PET-D-a
PET-D-b
PET-D-c
PET-C-NT
PET-C-a
PET-C-b
PET-C-c
PA6-C-NT
PA6-C-a
PA6-C-b
PA6-C-c
ABS-D-NT
ABS-D-a
ABS-D-b
ABS-D-c
MABS-02-D-NT
MABS-02-D-a
MABS-02-D-b
MABS-02-D-c
TCPS
PureCoat
Surface Code

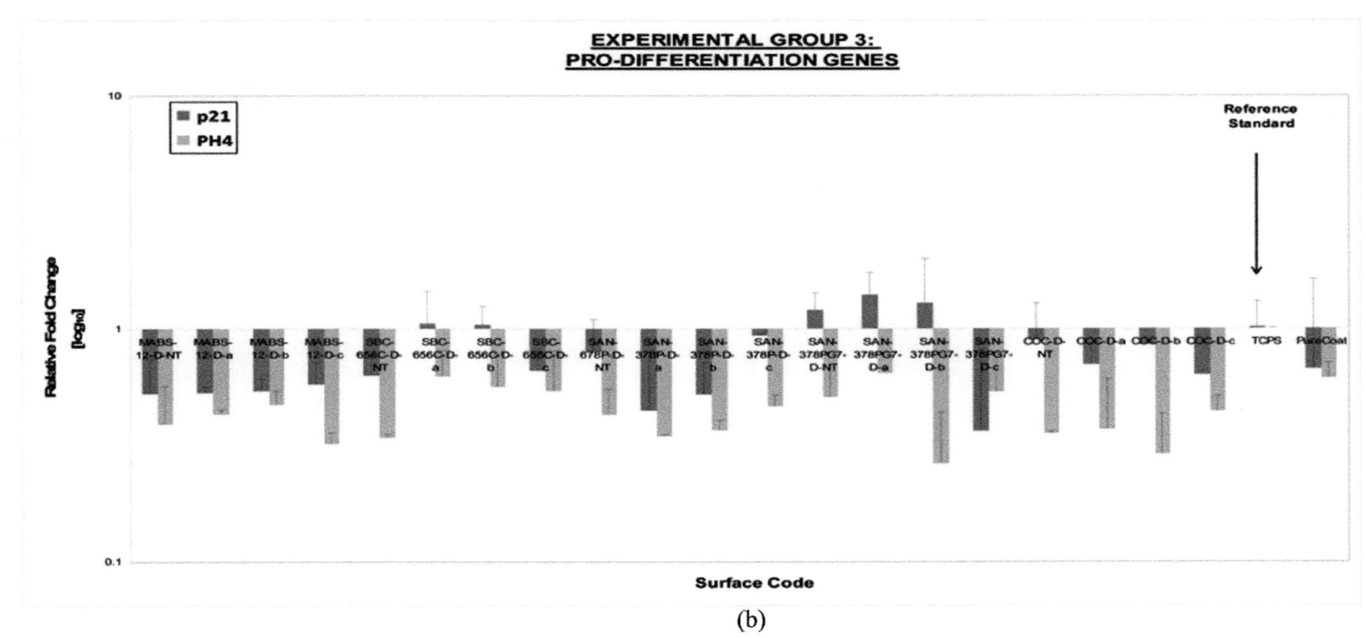

(b)

Figure 6.4 Effects of additional plastic surfaces on relative pro-stem cell and pro-differentiation gene expression in MSC cultures. Experiments were performed as in *Figure 6.3* but with additional surfaces examining the effects on gene expression related to (a) MSC maintenance (OCT4, NOTCH1, NANOG) or (b) differentiation (p21, PH4). Normalizations were performed with β-actin, and values were arbitrarily assigned a value of 1 relative to BD TCPS Falcon. BD PureCoat Amine plates also served as a reference material.

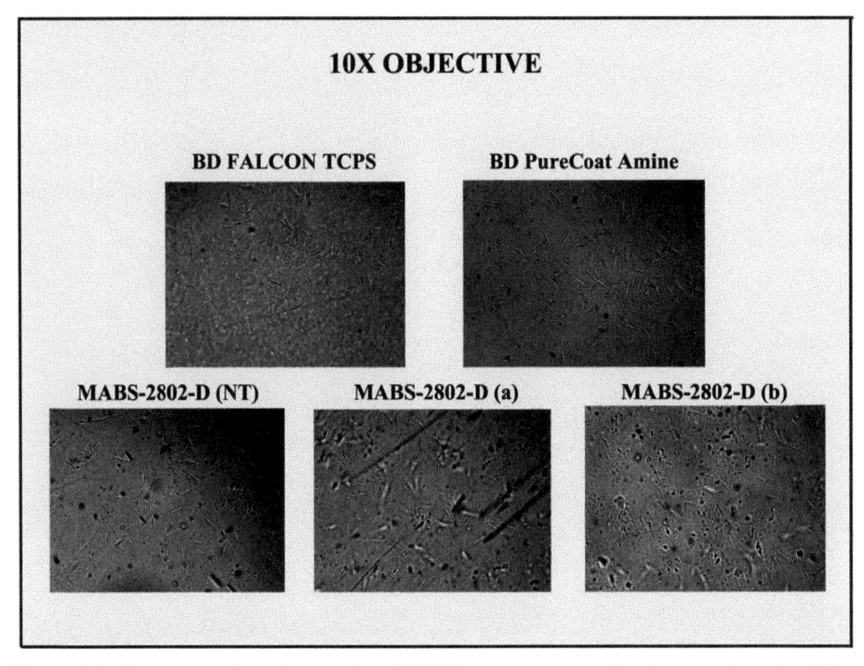

Figure 6.5 Representative pictographs of hMSC grown on control and test surfaces for 5 days. Images from cells grown on several non-opaque surfaces were included to show morphology on test coverslips.

currently the best available surface for expanding MSC. However, the use of TCPS is limited by changes in cell growth and function with extended culturing. Clinical applications require consistencies amongst the stem cells for extended periods of time. Part of the inconsistencies observed within labs culturing stem cells on these surfaces, may be due to the aging of the plasma-treated plates. Herein we studied the growth of hMSCs on several plastic surfaces in order to identify materials which outperform conventional TCPS, and that can potentially facilitate stem cell differentiation into specialized tissues.

Plasmas are often used to alter the surface properties of polymers and insert chemically reactive functionalities. The plasma discharge causes molecular fragmentations, bond fissions and ionizations generating reactive species and high energy photons that engage in subsequent reactions, resulting in cleaning, ablation, crosslinking, and surface chemical functionalization of the plasma-treated substrates. Typically noble gas plasmas (e.g., He or Ar) are effective in etching the surface whereas chemically reactive plasmas add new functional groups. The significant decrease in contact angle of our plasma treated coverslips is indicative of increased surface hydrophilicity. Under the experimental conditions used, the depth of plasma surface modification is about 20 Å. Select coverslip surfaces were analyzed by XPS to determine the elemental surface composition and extent of surface derivatization. Since the XPS depth of analysis is 100 Å, the data reported in Table 6.3 averages the elemental analysis of the derivatized surfaces with the bulk values. One limitation of XPS is that it cannot detect hydrogen, therefore the ratio of elements reported discounts the presence of hydrogen.

The XPS spectra are obtained by irradiating the material to be analyzed with a beam of X-rays while simultaneously measuring the kinetic energy and number of electrons that escape from the surface. The output data is the binding energy of the ejected electron which relates to the orbital from which the electron is ejected, characteristic of each element. The number of electrons detected with a specific binding energy is proportional to the number of corresponding atoms in the sample. This then provides the percent of each atom in the sample. The chemical environment and oxidation state of the atom can be determined through the shifts of the peaks within the range expected. If the electrons are shielded then it is easier, or requires less energy, to remove them from the atom, i.e., the binding energy is low. The corresponding peaks will shift to a lower energy in the expected range. If the core electrons are not shielded as much, such as the atom being in a high oxidation state, then just the opposite occurs.

With the exception of POM, all other plasma treatments enhanced the oxygen content of the plastics therefore oxidizing the surface by incorporation of oxygen-containing moieties (O/C ratio increases after treatment). For example, the XPS data of the coverslips made from SBC 656 C showed not only an increase of the elemental oxygen percentage at the surface after plasma treatment, but also an increase of the fraction of carbon atoms with higher oxidation state. The aliphatic hydrocarbon sp^3 carbon (1s) binding energy is 284.8 eV. Electronegative substituents decrease the electron density on the carbon atom causing small increases in the C(1s) binding energies. The experimental C(1s) spectra for SBC 656 C were resolved by including Gaussian contributions from higher binding energy peaks normally assigned to C(1s) substituted with oxygen. The nitrogen enrichment at the surface was the highest for plasma treatment (a), employing a gaseous mixture of ammonia and hydrogen. The binding energy of the N(1s) peak at 399 eV corresponds to

sp^3 nitrogen bonded to sp^3 hybridized carbon, and is good evidence for amine groups on the surface [28, 29]. The chemical state identifications based on the measured binding energies are presented in Table 6.4. The fitting of the XPS peaks is presented in Figure 6.6.

Plasma treatment suffers from a short-lived nature as treated chains could reptate into the polymer bulk with the surface reverting partly to the original untreated state. This may lead to significant variability as the surface chemical composition changes upon storage after plasma fabrication. The plasma conditions used herein are known to generate chemically-modified surfaces stable for at least 2 months. However, two months after plasma treatment all surfaces had their contact angle reversed to almost the initial values before treatment. Surface aging seems to depend on the polymer chain flexibility (T_g) and should be taken into account as they could contribute to product variability.

The coverslips made with molds having a matte finish had high roughness (RMS 167–434 nm). The coverslips made with molds having a glossy finish had very low surface roughness (RMS 6–18 nm), with the exception of SAN 378 PG-7 (RMS 235nm) (Figure 6.7). This plastic is a tough glass fiber-reinforced SAN (30% glass fiber) with the highest modulus in the series of 12 GPa.

All test surfaces were divided into three experimental groups which were analyzed independently. BD TCPS Falcon and BD PureCoat Amine plates served as reference standards. The effects of the BASF polymeric surfaces on stem cell maintenance and differentiation were screened by stem cell specific (OCT-4; NOTCH1; FOXD3) and differentiation specific (PH-4; p63) biomarkers. In analyzing the data, an increase in the stem cell specific or a decrease in the differentiation-specific genes indicates a "positive result" compared to reference standard. For the first experimental group, "positive results" in two of the three studied genes was viewed as warranting further analysis. For the second and third experimental groups, "positive results" in three of the five studied genes was viewed as warranting further analysis.

The following plastic surfaces represent positive hints: (a) in experimental group 1 surfaces POM G-a, POM G-b, and POM M1-c with better than reference standard in 2 of 3 stem cell-specific genes; (b) in experimental group 2 surfaces PA6 M1-a, PA6 M1-b, and ABS G-NT with 3 of 5 stem cell- or differentiation-specific genes better then reference; and (c) in experimental group 3 surfaces SBC 656 C G-c, SAN 378 PG-7 G-NT, SAN 378 PG-7 G-a, SAN-378PG7 G-b, and COC G-NT with 4 of 5 genes better than

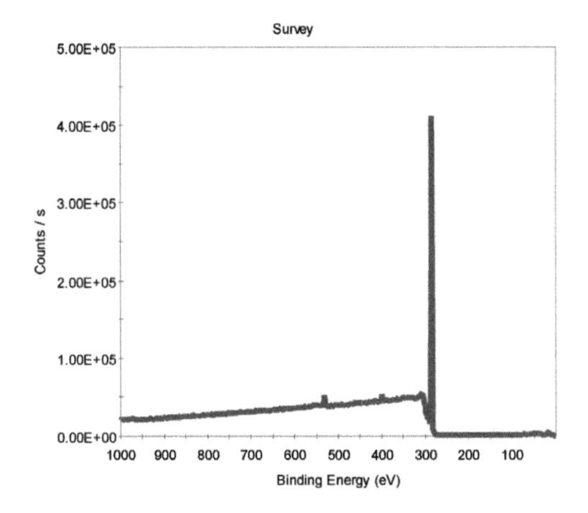
Survey
5.00E+05
4.00E+05
3.00E+05
2.00E+05
1.00E+05
0.00E+00
Counts / s
1000 900 800 700 600 500 400 300 200 100
Binding Energy (eV)

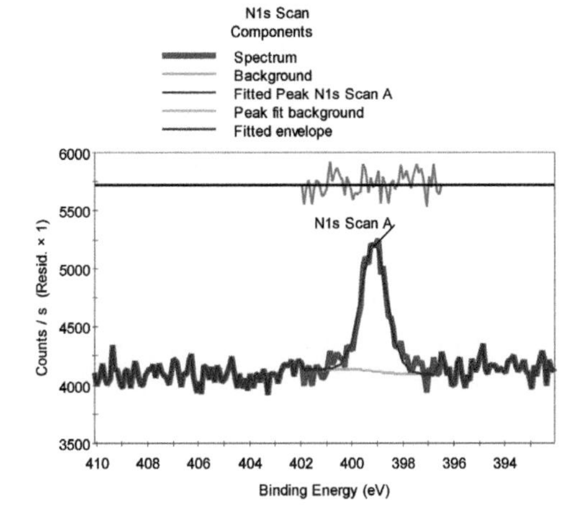
N1s Scan
Components
Spectrum
Background
Fitted Peak N1s Scan A
Peak fit background
Fitted envelope
6000
5500
5000
4500
4000
3500
Counts / s (Resid. × 1)
N1s Scan A
410 408 406 404 402 400 398 396 394
Binding Energy (eV)

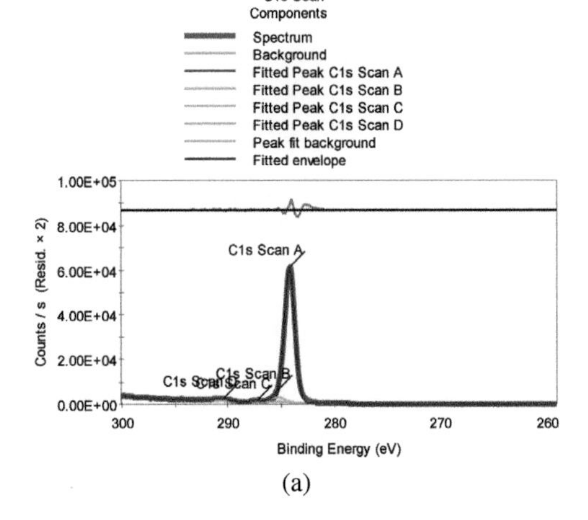
C1s Scan
Components
Spectrum
Background
Fitted Peak C1s Scan A
Fitted Peak C1s Scan B
Fitted Peak C1s Scan C
Fitted Peak C1s Scan D
Peak fit background
Fitted envelope
1.00E+05
8.00E+04
6.00E+04
4.00E+04
2.00E+04
0.00E+00
Counts / s (Resid. × 2)
C1s Scan A
C1s Scan B
C1s Scan C
300 290 280 270 260
Binding Energy (eV)

(a)

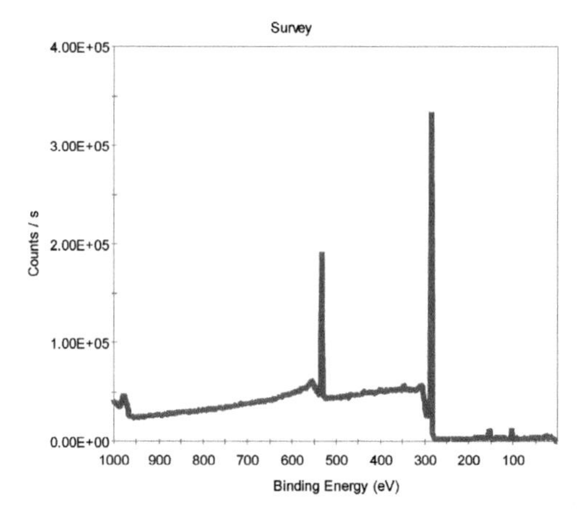
Survey
Counts / s
Binding Energy (eV)

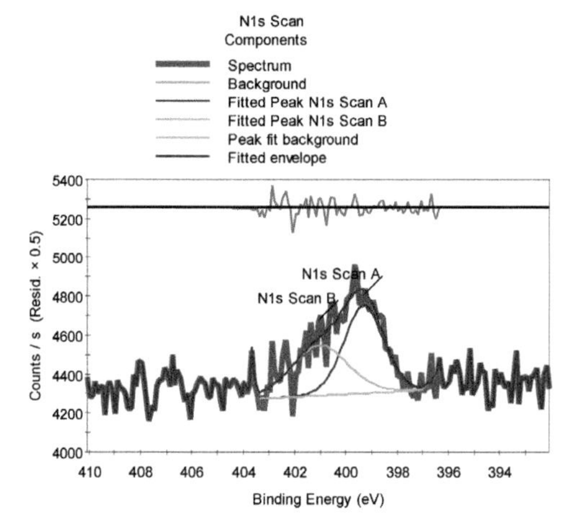
N1s Scan
Components
Spectrum
Background
Fitted Peak N1s Scan A
Fitted Peak N1s Scan B
Peak fit background
Fitted envelope
N1s Scan A
N1s Scan B
Counts / s (Resid. × 0.5)
Binding Energy (eV)

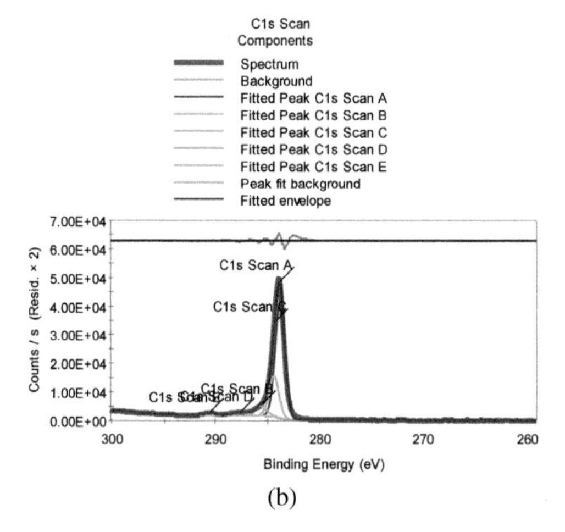
C1s Scan
Components
Spectrum
Background
Fitted Peak C1s Scan A
Fitted Peak C1s Scan B
Fitted Peak C1s Scan C
Fitted Peak C1s Scan D
Fitted Peak C1s Scan E
Peak fit background
Fitted envelope
C1s Scan A
Counts / s (Resid. × 2)
Binding Energy (eV)

(b)

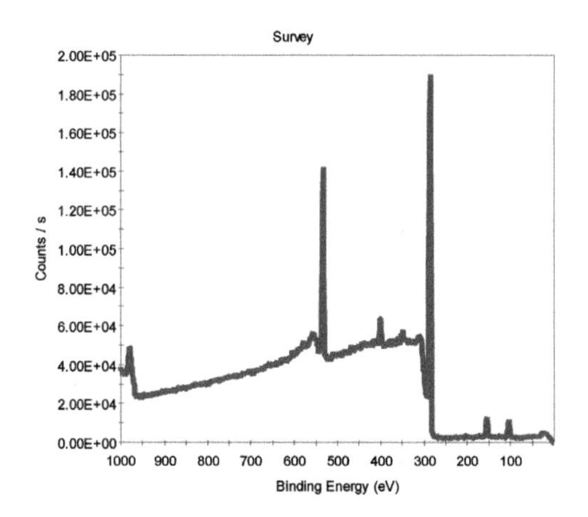
Survey
Counts / s
Binding Energy (eV)

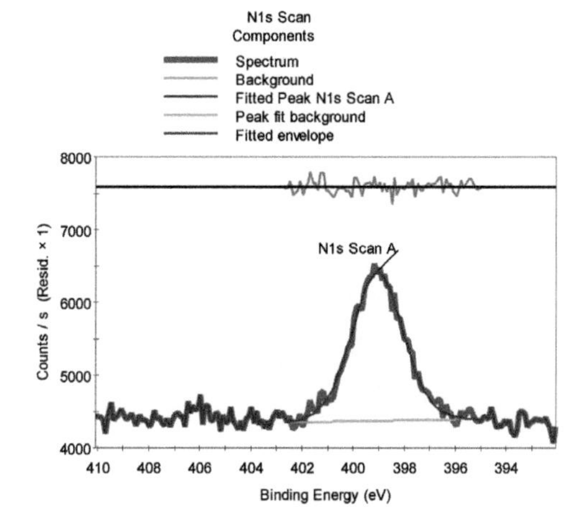
N1s Scan
Components
Spectrum
Background
Fitted Peak N1s Scan A
Peak fit background
Fitted envelope
N1s Scan A
Counts / s (Resid. × 1)
Binding Energy (eV)

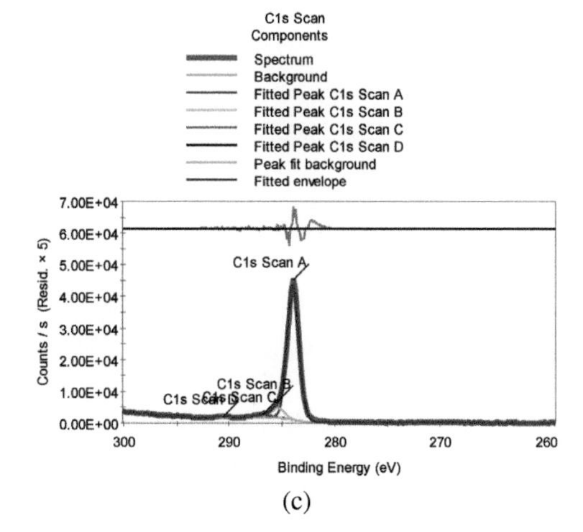
C1s Scan
Components
Spectrum
Background
Fitted Peak C1s Scan A
Fitted Peak C1s Scan B
Fitted Peak C1s Scan C
Fitted Peak C1s Scan D
Peak fit background
Fitted envelope
C1s Scan A
C1s Scan B
C1s Scan C
Counts / s (Resid. × 5)
Binding Energy (eV)

(c)

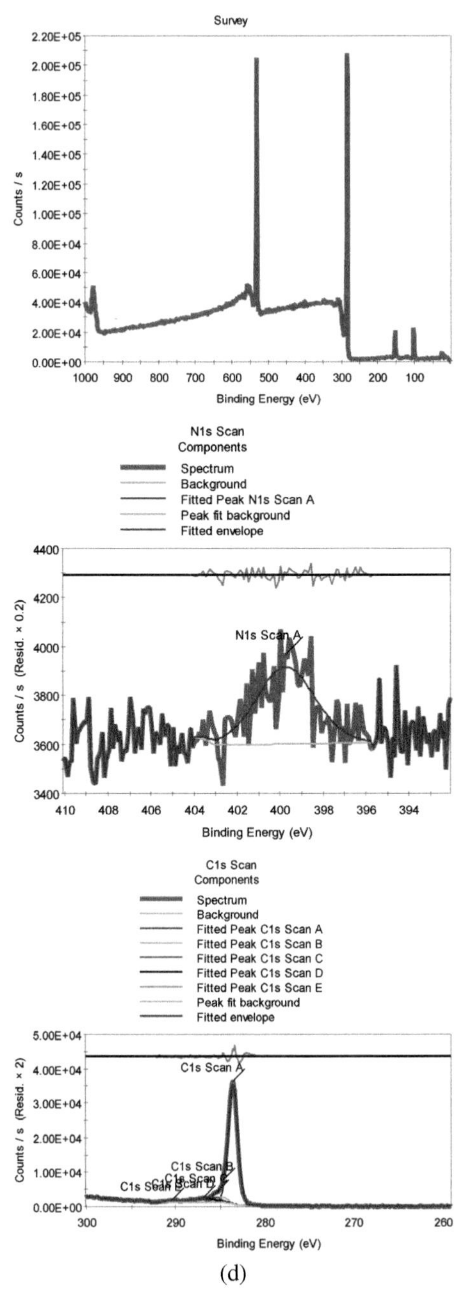

(d)

Figure 6.6 XPS spectra, C(1s), and N(1s) peak fittings of SBC 656 C coverslips: (a) SBC 656 C G-NT; (b) SBC 656 C G-a; (c) SBC 656 C G-b; and (d) SBC 656 C G-c. The peaks were fit with Gaussian line-shapes after background subtraction.

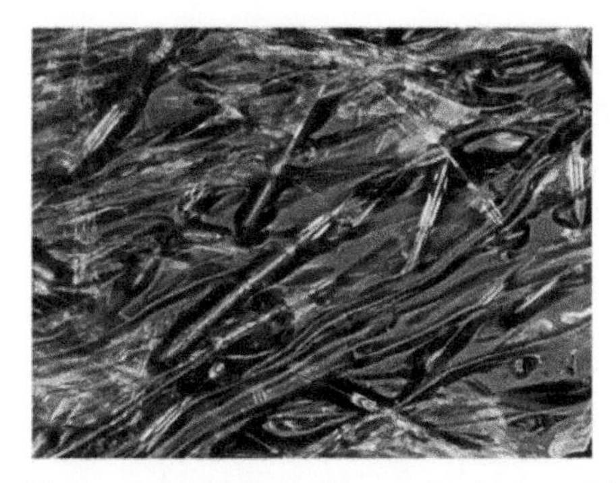

Figure 6.7 Optical micrograph pictures of SAN 378 PG-7 (G-c) coverslips. The pictures were taken with a Nikon Inverted Metallurgical Epiphot 200 in the darkfield reflective mode.

reference, and surface SAN 378 PG-7 G with 5 of 5 genes better than reference. Of the many surface physical characteristics the roughness and fibrous structure of the glass fiber-reinforced poly(styrene-acrylonitrile) SAN 378 PG-7 surface had the most prevalent effect on facilitating MSC expansion with preservation of stem cell function.

6.5 Conclusions

We tested a panel of plastic surfaces with different physical characteristics in order to identify materials which outperform conventional tissue culture polystyrene (TCPS) in their ability to expand and maintain hMSCs in an undifferentiated state. Several surfaces exhibited a statistically significant increase in the stem cell specific-genes and/or a decrease in the differentiation-specific genes, indicating a "positive result" compared to the TCPS reference standard. The plastics with positive results did not correlate with a specific surface chemistry or plasma treatment. Of the many surface physical characteristics, the roughness and fibrous structure of a glass fiber-reinforced poly(styrene-acrylonitrile) polymeric surface had the most prevalent effect on facilitating hMSC expansion with preservation of stem cell function. These findings highlight, in particular, the important role of the surface mechanical properties in cell/material interactions. Future investigations are planned to validate these results across multiple cell culture passages, involving the effects on cell phenotype, long-term morphology, protein expression, and functional assessment.

Acknowledgements

The authors would like to acknowledge Dr. Nancy Brungard and Melissa Thornton from BASF, Iselin, NJ, for performing the XPS experiments, and Dr. Rachel Dong and Corola Jernigan from BASF, Tarrytown, NY, for the AFM and optical microscope analysis.

References

[1] Yan, Y., et al., *Directed differentiation of dopaminergic neuronal subtypes from human embryonic stem cells.* Stem cells, 2005. **23**(6): 781–90.

[2] Kim, N.R., et al., *Discovery of a new and efficient small molecule for neuronal differentiation from mesenchymal stem cell.* Journal of medicinal chemistry, 2009. **52**(24): 7931–3.

[3] Totey, S. and R. Pal, *Adult stem cells: a clinical update.* Journal of stem cells, 2009. **4**(2): 105–21.

[4] Trzaska, K.A., et al., *Brain-derived neurotrophic factor facilitates maturation of mesenchymal stem cell-derived dopamine progenitors to functional neurons.* Journal of neurochemistry, 2009. **110**(3): 1058–69.

[5] Momin, E.N., et al., *Mesenchymal stem cells: new approaches for the treatment of neurological diseases.* Current stem cell research & therapy, 2010. **5**(4): 326–44.

[6] Patel, N., et al., *Developmental regulation of TAC1 in peptidergic-induced human mesenchymal stem cells: implication for spinal cord injury in zebrafish.* Stem cells and development, 2012. **21**(2): 308–20.

[7] Campagnoli, C., et al., *Identification of mesenchymal stem/progenitor cells in human first-trimester fetal blood, liver, and bone marrow.* Blood, 2001. **98**(8): 2396–402.

[8] Castillo, M., et al., *The immune properties of mesenchymal stem cells.* International journal of biomedical science: IJBS, 2007. **3**(2): 76–80.

[9] Dominici, M., et al., *Heterogeneity of multipotent mesenchymal stromal cells: from stromal cells to stem cells and vice versa.* Transplantation, 2009. **87**(9 Suppl): S36–42.

[10] Cho, K.J., et al., *Neurons derived from human mesenchymal stem cells show synaptic transmission and can be induced to produce the neurotransmitter substance P by interleukin-1 alpha.* Stem cells, 2005. **23**(3): 383–91.

[11] Greco, S.J., et al., *An interdisciplinary approach and characterization of neuronal cells transdifferentiated from human mesenchymal stem cells.* Stem cells and development, 2007. **16**(5): 811–26.

[12] Greco, S.J., et al., *Synergy between the RE-1 silencer of transcription and NFkappaB in the repression of the neurotransmitter gene TAC1 in human mesenchymal stem cells.* The Journal of biological chemistry, 2007. **282**(41): 30039–50.

[13] Cho, J., P. Rameshwar, and J. Sadoshima, *Distinct roles of glycogen synthase kinase (GSK)-3alpha and GSK-3beta in mediating cardiomyocyte differentiation in murine bone marrow-derived mesenchymal stem cells.* The Journal of biological chemistry, 2009. **284**(52): 36647–58.

[14] Potian, J.A., et al., *Veto-like activity of mesenchymal stem cells: functional discrimination between cellular responses to alloantigens and recall antigens.* Journal of immunology, 2003. **171**(7): 3426–34.

[15] Chan, J.L., et al., *Antigen-presenting property of mesenchymal stem cells occurs during a narrow window at low levels of interferon-gamma.* Blood, 2006. **107**(12): 4817–24.

[16] Romieu-Mourez, R., et al., *Regulation of MHC class II expression and antigen processing in murine and human mesenchymal stromal cells by IFN-gamma, TGF-beta, and cell density.* Journal of immunology, 2007. **179**(3): 1549–58.

[17] Le Blanc, K., et al., *Mesenchymal stem cells for treatment of steroid-resistant, severe, acute graft-versus-host disease: a phase II study.* Lancet, 2008. **371**(9624): 1579–86.

[18] Stagg, J., *Immune regulation by mesenchymal stem cells: two sides to the coin.* Tissue antigens, 2007. **69**(1): 1–9.

[19] Greco, S.J. and P. Rameshwar, *Microenvironmental considerations in the application of human mesenchymal stem cells in regenerative therapies.* Biologics: targets & therapy, 2008. **2**(4): 699–705.

[20] Buron, F., et al., *Human mesenchymal stem cells and immunosuppressive drug interactions in allogeneic responses: an in vitro study using human cells.* Transplantation proceedings, 2009. **41**(8): 3347–52.

[21] Wang, Y., et al., *Bone marrow-derived mesenchymal stem cells inhibit acute rejection of rat liver allografts in association with regulatory T-cell expansion.* Transplantation proceedings, 2009. **41**(10): 4352–6.

[22] Tao, X.R., et al., *Clonal mesenchymal stem cells derived from human bone marrow can differentiate into hepatocyte-like cells in injured livers of SCID mice.* Journal of cellular biochemistry, 2009. **108**(3): 693–704.

[23] Mohseny, A.B., et al., *Osteosarcoma originates from mesenchymal stem cells in consequence of aneuploidization and genomic loss of Cdkn2.* The Journal of pathology, 2009. **219**(3): 294–305.

[24] Molcanyi, M., et al., *Developmental potential of the murine embryonic stem cells transplanted into the healthy rat brain–novel insights into tumorigenesis.* Cellular physiology and biochemistry: international journal of experimental cellular physiology, biochemistry, and pharmacology, 2009. **24**(1–2): 87–94.

[25] Shyong Siow, K., et al., *Plasma methods for the generation of chemically reactive surfaces for biomolecule immobilization and cell colonization – A review.* Plasma processes and polymers, 2006. **3**: 392–418.

[26] Hartwig, A. et al., *Smolders surface amination of poly(acrylonitrile).* Advances in colloid and interface science, 1994. **52**: 5–78.

[27] Klages, C.P., et al., *Atmospheric-pressure plasma amination of polymer surfaces.* Journal of adhesion science and technology, 2010. **24**: 1167–1180.

[28] Gammona, W.J., et al., *Experimental comparison of N(1s) X-ray photoelectron spectroscopy binding energies of hard and elastic amorphous carbon nitride films with reference organic compounds.* Carbon, 2003. **41**: 1917–1923.

[29] Nagatsu, M., et al., *Functionalization of polymer surfaces using microwave plasma chemical modification.* Journal of photopolymer science and technology, 2008. **21**(2): 257–261.

7

Embryonic Stem Cell Markers in Cancer: Cripto-1 Expression in Glioblastoma

Meg Duroux[1], Linda Pilgaard[1] and Pia Olsen[1,2]

[1]Laboratory for Cancer Biology, Institute of Health Science and Technology, Aalborg University, Denmark
[2]Aalborg University Hospital, Department for Neuro Surgery, Denmark

Abstract

Human glioblastoma multiforme (GBM) is an aggressive form of brain tumor with a very poor prognosis. Heterogeneous in nature, the tumors characteristically grow by infiltrating the surrounding brain tissue and relapse is inevitable. The poor prognosis of GBM justifies the search to identify novel candidates for prognostic markers and therapeutic targeting. The pool of known embryonic and stem cell markers provide a resource of potential targets to investigate. Exploring the re-emergence of the embryonic marker Teratocarcinoma-Derived Growth Factor (TDGF-1) also known as Cripto-1 (CR-1) in GBM tissue and blood has provided further insight into the regulatory mechanisms governing GBM pathology. In GBM patients, high CR-1 protein levels in blood plasma significantly correlate with a shorter overall survival and survival per-se is linked to high CR-1 levels in tissue of younger patients. CR-1 expression is localized to different areas of tumor tissue, i.e., in the malignant cells in zones of proliferation, in the vicinity of endothelial cells, microvasculature and in some areas co-localization with the stem cell marker SOX-2. With these new findings, CR-1 could be a prognostic biomarker for GBM in tissue and blood with the potential of being a therapeutic target. Here, we provide an update on the expression of CR-1 in GBM and reflect on the future perspectives of these discoveries.

Keywords: Glioblastoma, Cancer stem cell marker, Cripto-1, Invasion, Hypoxia.

7.1 Introduction

Glioblastoma multiform (GBM) brain tumors are characterized by their highly invasive growth and excessive formation of new and abnormal vasculature [1]. During the last decade, new therapeutic strategies and better diagnostics have improved cancer survival in general. However, the prognosis for GBM remains poor. Even with multimodal treatment consisting of radical surgery, chemo- and radiotherapy, the tumor niche facilitates tumor regrowth and relapse is inevitable [2]. The 5 year survival rate for GBM patients is merely 10% (Danish Neuro Oncology Group, DNOG 2014). The formation of abnormal vasculature and the migration of GBM tumor cells are thought to be the cause of GBM resistance to therapy [1, 3–5]. Hence, in the search for more efficient therapies, the focus has moved towards targeting the rogue population of cells that are referred to as tumor initiating cells or cancer stem cells (CSC). Different populations of GBM cancer stem cells (GSCs) sharing expression patterns with embryonic stem cell markers have been identified. For instance, Oct-4, SOX-2, Nanog are genes involved in stem cell maintenance and these have been shown to be up-regulated in GBM [6, 7].

7.2 Glioma Stem Cells (GSC)

Classically, GSCs share properties of normal stem cells, importantly, the ability of self-renewal and unlimited proliferation both *in vitro* and *in vivo*. *In vitro*, these properties have been investigated under stem cell promoting growth conditions and it has been shown that the GSCs are capable of establishing new neurospheres even after many passages [4, 8]. GSC's are typically resistant to radiation therapy and chemotherapy, because of their preferential response of the DNA damage checkpoint with enhanced DNA repair capacity [9]. They are often found to reside in perivascular niches, and it has been shown that glioma cells expressing CSC markers such as CD133, HIF2α and CD171 are localized near blood vessels, indicating that these niches may provide a specific microenvironment for the maintenance of GSC population [9]. The hypoxic niche plays a critical role in maintaining GSC, and the extensive vascular network, a hallmark of GBM [10–14]. Recent studies have highlighted the transdifferentiation potential of pericyte-like cancer cells that could in turn participate to the cellular heterogeneity found in GBM [15].

The identification and subsequent targeting of novel GSCs markers could be an important step towards developing new treatments of GBM and hence minimizing recurrence. To date, a number of markers have been identified for GBM stem cell. Some of the most commonly used CSC markers include CD133, SOX-2, Nanog, OCT3/4 and Nestin [5, 16, 17]. The marker CD133, often referred to as Prominin 1 has been used quite extensively as a GSC biomarker. CD133-positive cells isolated from GBM elicited stem cell characteristics *in vitro* and were able to form tumors when grown *in vivo* [4, 5]. Contrary to the research by Singh et al., CD133-negative cells derived from fresh GBM tumor tissue, along with established GBM cell lines, also have tumorigenic properties *in vivo* [18–20]. The complexity of the protein and the inconsistencies with immunostaining has gradually moved the scientific population away from using this as a canonical GSC marker [5, 18, 21]. In light of these contradictory findings, new GSC biomarkers are essential for a more complete characterization of the GSC population that could lead to therapeutic targeting of GSC's in GBM, and hereby improve the prognosis of the disease.

7.3 A New Cancer Stem Cell Marker in the Tumor Scaffold

We know that the re-emergence of embryonic signaling pathways plays a key role in cancer biology [22]. Through extensive literature search and analogy to cancer stem cell like characteristics in other cancer cell types, a promising candidate as a novel GSC marker was identified [23, 24]. Human Cripto-1 (CR-1, also known as teratocarcinoma-derived growth factor-1), is the founding member of the epidermal growth-factor (EGF)-Cripto-1-FRL-1-Cryptic (CFC) family [25, 26]. CR-1 has a key role in a range of processes such as migration, angiogenesis, and maintenance of undifferentiated stem cells [26–31]. In development, CR-1 is involved in the highly coordinated epithelial-mesenchymal transition converting densely packed immobile epithelial cells into invasive mesenchymal cells [32–34]. CR-1 is expressed at low levels in normal adult tissues and has been shown to be up-regulated in several solid cancers [35–38]. In a number of recent studies, abnormal and high levels of CR-1 have been shown to correlate to malignant transformation manifested as tumor invasiveness, metastatic spreading and resulting poor prognosis [35, 37, 38]. CR-1 targeted therapies are being developed and are promising for a number of solid tumors. CR-1 expression

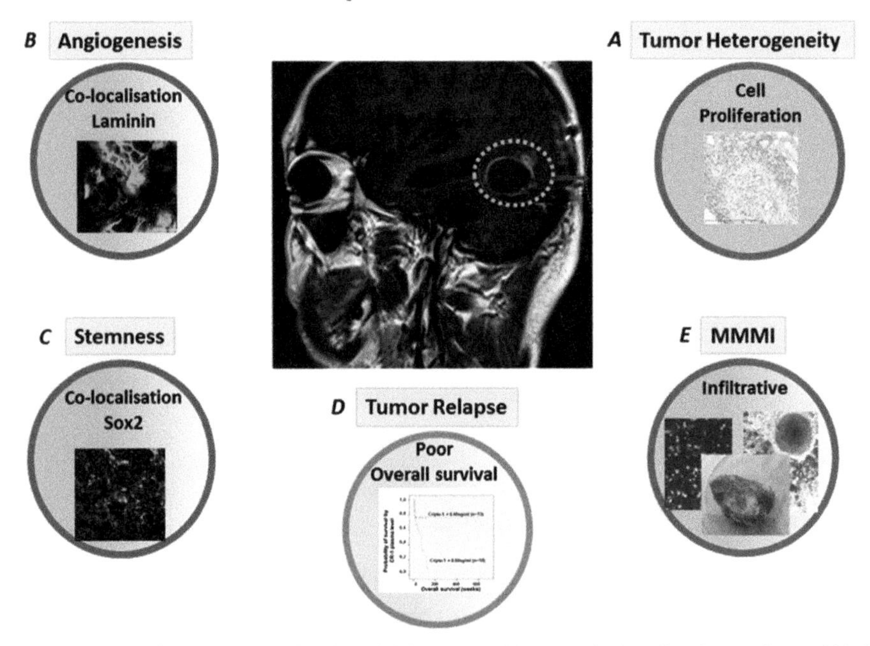

Figure 7.1 Cripto-1 expression in glioblastoma. CR-1 can be localized to regions of high proliferation and around glomeruloid vasculature (A). CR-1 expression in and around the vasculature marked by staining of laminin (B). Co-localization with other stemness markers like SOX-2 (C). High CR-1 expression in blood correlates to shorter survival (D). CR-1 is linked to the infiltrative behavior defined as Mesenchymal Mode of Migration and Invasion (MMMI) in primary and relapse tumors (E).

and its clinical relevance have not been extensively investigated in GBM. However, our recent findings, combined with the comprehensive study by Tysnes et al., 2013 have highlighted the potential of CR-1 as a "CSC like" marker in GBM pathology, summarized in Figure 7.1.

7.4 Cr-1 Expression in GBM Tissue and the Angiogenic Phenotype

CR-1 was found to be expressed in GBM patients at both the mRNA and protein level [23, 24]. In patient tissue and in xenograft tumors derived from GBM model cell lines, CR-1 positive cells were found to reside in distinct niches [24]. In patient derived tumor tissue, CR-1 was localized to some degree around the glomerular like structures in the palisading layers and around the blood vessels [24]. This niche dependent CR-1 expression could be attributed to the hypoxic response based on the biological heterogeneity found in the tumor. In model cell lines, this notion is supported as CR-1 has been seen to be highly up-regulated in xenografted cultures forming tumors compared to *in vitro* cultures. Here the expression co-localized with the endothelial basal membrane protein laminin and the stem cell marker SOX-2 (unpublished data, Pilgaard et al.). An inducing effect of Cr-1 on proliferation, migration, invasion and its role as an angiogenic factor has been well demonstrated in endothelial cells, carcinoma and in melanoma [31, 39, 40]. Building on these facts, the inherent hypoxia induced expression seen in other studies may contribute to the tumor vascularization in GBM [41]. To be able to account for the degree of tumor vascularization and correlate this to the level of CR-1 expression and to pinpoint the cell type of origin, would further our understanding about the possible role and contribution of CR-1 to the GBM angiogenic phenotype.

7.5 CR-1 Expression Linked to Poorer Prognosis and Shorter Survival

When looking at the expression level of CR-1 in primary GBM tissue and in plasma, a correlation to the clinical outcome was shown. It was found that higher CR-1 expression levels detected with immunohistochemistry were associated with significantly shorter survival in a subset of younger GBM patients [23]. Similarly, in our study, the protein levels found in tissue were covering quite a broad range of CR-1 expression when measured with ELISA. This could depict the disease progression, or rather the highly heterogenous nature of the tumor. However, when analyzing the plasma CR-1 levels, this was correlated with overall survival of the patients when performing a Kaplan Meier Cox Regression Analysis. High CR-1 was shown to be associated with shorter overall survival [24].

7.6 Conclusion

The higher expression of CR-1 in GBM vasculature and correlation with survival support the notion that CR-1 is an important requisite for GBM progression.

References

[1] Plate, K.H. and W. Risau, *Angiogenesis in malignant gliomas.* Glia, 1995. 15(3): 339–47.

[2] Stupp, R., et al., *Radiotherapy plus concomitant and adjuvant temozolomide for glioblastoma.* N Engl J Med, 2005. 352(10): 987–96.

[3] Kang, M.K. and S.K. Kang, *Tumorigenesis of chemotherapeutic drug-resistant cancer stem-like cells in brain glioma.* Stem Cells Dev, 2007. 16(5): 837–47.

[4] Singh, S.K., et al., *Identification of a cancer stem cell in human brain tumors.* Cancer Res, 2003. 63(18): 5821–8.

[5] Singh, S.K., et al., *Identification of human brain tumour initiating cells.* Nature, 2004. 432(7015): 396–401.

[6] Guo, Y., et al., *Expression profile of embryonic stem cell-associated genes Oct 4, SOX-2 and Nanog in human gliomas.* Histopathology. 59(4): 763–75.

[7] Huang, Z., et al., *Cancer stem cells in glioblastoma-molecular signaling and therapeutic targeting.* Protein Cell, 2010. 1(7): 638–55.

[8] McCord, A.M., et al., *Physiologic oxygen concentration enhances the stem-like properties of CD133+ human glioblastoma cells in vitro.* Mol Cancer Res, 2009. 7(4): 489–97.

[9] Huang, Z., et al., *Cancer stem cells in glioblastoma–molecular signaling and therapeutic targeting.* Protein Cell. 1(7): 638–55.

[10] Calabrese, C., et al., *A perivascular niche for brain tumor stem cells.* Cancer Cell, 2007. 11(1): 69–82.

[11] Birner, P., et al., *Vascular patterns in glioblastoma influence clinical outcome and associate with variable expression of angiogenic proteins: evidence for distinct angiogenic subtypes.* Brain Pathol, 2003. 13(2): 133–43.

[12] Heddleston, J.M., et al., *The hypoxic microenvironment maintains glioblastoma stem cells and promotes reprogramming towards a cancer stem cell phenotype.* Cell Cycle, 2009. 8(20): 3274–84.

[13] Seidel, S., et al., *A hypoxic niche regulates glioblastoma stem cells through hypoxia inducible factor 2 alpha.* Brain, 2010. 133(Pt 4): 983–95.

[14] Hendriksen, E.M., et al., *Angiogenesis, hypoxia and VEGF expression during tumour growth in a human xenograft tumour model.* Microvasc Res, 2009. 77(2): 96–103.

[15] Appaix, F., et al., *Brain mesenchymal stem cells: The other stem cells of the brain?* World J Stem Cells. 6(2): 134–143.

[16] Tomuleasa, C., et al., *Functional and molecular characterization of glioblastoma multiforme-derived cancer stem cells.* J BUON. 15(3): 583–91.

[17] Berezovsky, A.D., et al., *SOX-2 promotes malignancy in glioblastoma by regulating plasticity and astrocytic differentiation.* Neoplasia. 16(3): 193–206 e25.

[18] Beier, D., et al., *CD133(+) and CD133(−) glioblastoma-derived cancer stem cells show differential growth characteristics and molecular profiles.* Cancer Res, 2007. 67(9): 4010–5.

[19] Clement, V., et al., *Limits of CD133 as a marker of glioma self-renewing cells.* Int J Cancer, 2009. 125(1): 244–8.

[20] Shmelkov, S.V., et al., *CD133 expression is not restricted to stem cells, and both CD133+ and CD133− metastatic colon cancer cells initiate tumors.* J Clin Invest, 2008. 118(6): 2111–20.

[21] Hermansen, S.K., et al., *Inconsistent immunohistochemical expression patterns of four different CD133 antibody clones in glioblastoma.* J Histochem Cytochem. 59(4): 391–407.

[22] Ben-Porath, I., et al., *An embryonic stem cell-like gene expression signature in poorly differentiated aggressive human tumors.* Nat Genet, 2008. 40(5): 499–507.

[23] Tysnes, B.B., et al., *Age-Dependent Association between Protein Expression of the Embryonic Stem Cell Marker Cripto-1 and Survival of Glioblastoma Patients.* Transl Oncol, 2013. 6(6): 732–41.

[24] Pilgaard, L., et al., *Cripto-1 expression in glioblastoma multiforme.* Brain Pathol, 2014.

[25] Brandt, R., et al., *Identification and biological characterization of an epidermal growth factor-related protein: cripto-1.* J Biol Chem, 1994. 269(25): 17320–8.

[26] Ciccodicola, A., et al., *Molecular characterization of a gene of the 'EGF family' expressed in undifferentiated human NTERA2 teratocarcinoma cells.* EMBO J, 1989. 8(7): 1987–91.

[27] Ding, J., et al., *Cripto is required for correct orientation of the anterior-posterior axis in the mouse embryo.* Nature, 1998. 395(6703): 702–7.

[28] Xu, C., et al., *Abrogation of the Cripto gene in mouse leads to failure of postgastrulation morphogenesis and lack of differentiation of cardiomyocytes.* Development, 1999. 126(3): 483–94.

[29] Dono, R., et al., *The murine cripto gene: expression during mesoderm induction and early heart morphogenesis.* Development, 1993. 118(4): 1157–68.
[30] Johnson, S.E., J.L. Rothstein, and B.B. Knowles, *Expression of epidermal growth factor family gene members in early mouse development.* Dev Dyn, 1994. 201(3): 216–26.
[31] Bianco, C., et al., *Role of human cripto-1 in tumor angiogenesis.* J Natl Cancer Inst, 2005. 97(2): 132–41.
[32] Zhong, X.Y., et al., *Positive association of up-regulated Cripto-1 and down-regulated E-cadherin with tumour progression and poor prognosis in gastric cancer.* Histopathology, 2008. 52(5): 560–8.
[33] Hay, E.D., *An overview of epithelio-mesenchymal transformation.* Acta Anat (Basel), 1995. 154(1): 8–20.
[34] Viebahn, C., *Epithelio-mesenchymal transformation during formation of the mesoderm in the mammalian embryo.* Acta Anat (Basel), 1995. 154(1): 79–97.
[35] Fuchs, I.B., et al., *The prognostic significance of epithelial-mesenchymal transition in breast cancer.* Anticancer Res, 2002. 22(6A): 3415–9.
[36] Gong, Y.P., et al., *Overexpression of Cripto and its prognostic significance in breast cancer: a study with long-term survival.* Eur J Surg Oncol, 2007. 33(4): 438–43.
[37] Wechselberger, C., et al., *Cripto-1 enhances migration and branching morphogenesis of mouse mammary epithelial cells.* Exp Cell Res, 2001. 266(1): 95–105.
[38] Xue, C., et al., *The gatekeeper effect of epithelial-mesenchymal transition regulates the frequency of breast cancer metastasis.* Cancer Res, 2003. 63(12): 3386–94.
[39] Strizzi, L., et al., *Cripto-1: a multifunctional modulator during embryogenesis and oncogenesis.* Oncogene, 2005. 24(37): 5731–41.
[40] De Luca, A., et al., *Expression and functional role of CRIPTO-1 in cutaneous melanoma.* Br J Cancer, 2011. 105(7): 1030–8.
[41] Bianco, C., et al., *Cripto-1 is required for hypoxia to induce cardiac differentiation of mouse embryonic stem cells.* Am J Pathol, 2009. 175(5): 2146–58.

8

Artificial Corneas, and Reinforced Composite Implants for High Risk Donor Cornea Transplantation

May Griffith[1], Chyan-Jang Lee[1] and Oleksiy Buznyk[1,2]

[1]Integrative Regenerative Medicine Centre, and Department of Clinical and Experimental Medicine, Linköping University, Linköping, Sweden
[2]Department of Eye Burns, Ophthalmic Reconstructive Surgery, Keratoplasty & Keratoprosthesis, Filatov Institute of Eye Diseases and Tissue Therapy, Odessa, Ukraine

Abstract

Here, we review examples of artificial corneas that have been developed as alternatives to donor cornea transplantation. These consist of artificial corneas developed as prostheses and regenerative scaffolds. Examples of reinforced and composite implants developed within our group are profiled.

Keywords: Cornea, Regeneration, Biomaterials, Implants, HSV-1.

8.1 Introduction

8.1.1 Cornea Transplantation

In Sweden, as in most countries of the world, there is a serious shortage of donor organs for transplantation, and when organs are available, there is still a problem of immune rejection. The myth is that the cornea is an immune privileged site and hence transplantation is problem-free. However, in reality, there are many pathological conditions where the immune privilege is gone (e.g., chemical burns, severe/persistent viral and bacterial infections, severe dry eye, neuropathic issues, autoimmune conditions, etc.). In addition, while the two year success rates for cornea allografting are high – 85% in a developed EU nation like Sweden [1], the long-term success at 10–15 years falls to 55% [2], which is even lower than that for kidney transplantation [3]. In a developing country, however, even the short term rates start out low, at 69% in South India [4]. This is most likely due to the higher incidence of more severe pathologies such as chemical burns or inflammation. In the cornea, as in other solid organs, graft rejection is directed against the foreign cells, and this is managed by steroid use for 6 to 12 months, or even 2–3 years.

Worldwide, an estimated 10 million individuals are in need of corneal transplantation, but there is a severe shortage of good quality donor corneas. Disease transmission is pre-empted by very rigorous and expensive screening but still transmission of diseases such as HSV, rabies and Creutzfeldt-Jakob disease have been traced back to the donor corneas.

8.2 Artificial Corneas as Prostheses and Regeneration Templates

8.2.1 Artificial Corneas as Prostheses

In the cornea transplantation, the use of allogeneic donor corneas has remained the ‘gold standard’ or state-of-the-art for over a century despite numerous efforts to develop synthetic prostheses (called keratoprostheses or KPros). There have been many attempts to produce the optimal biomaterials to fabricate KPros. Most have resulted in a high proportion of rejection and need sustained immunosuppression but several are very successful and are the only options for corneas that are very badly damaged or diseased. The most successful prostheses and amongst the best known prostheses, the Boston KPro and Osteo-odonto-keratoprosthesis (OOKP) both have a biological interface. The former uses a human corneal rim as an interface while the latter uses a layer of oral mucosal cells that is wrapped around the tooth or bone implant that holds the acrylic optic. Another successful KPro that has been used since 1978 is the Filatov Institute K-Pro, which is usually implanted in leukomas that have been previously strengthened with oral mucosa or human donor corneas (Figure 8.1a–b). This device consists of an acrylic optic with tantalum support (Figure 8.1c). A cartilage graft taken from the ear or human donor corneal allograft consisting of posterior stromal layers and Descemet’s membrane is implanted together with the Filatov Institute K-Pro to enhance stability of the device in corneal layers [5].

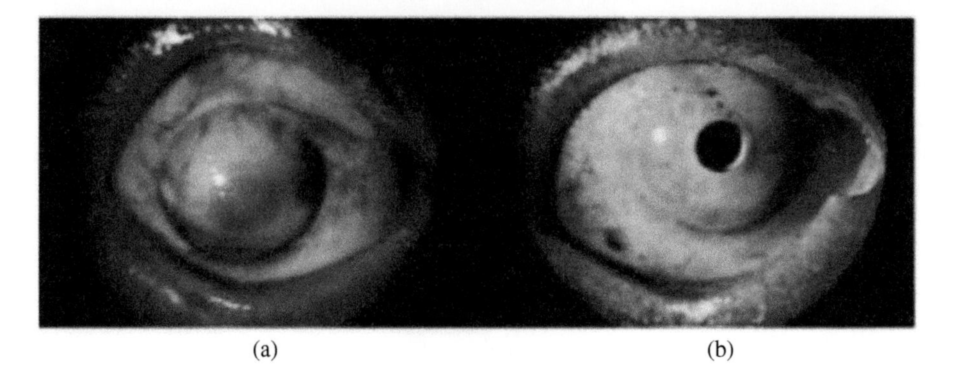

(a) (b)

(c)

Figure 8.1 Example of Filatov Institute KPro used in a patient with avascular leukoma. a) Preoperatively, the cornea was in the terminal stage of bullous keratopathy, aphakia, having been previously operated on for secondary glaucoma; b) After keratoprosthesis implantation, there was simultaneous intracorneal strengthening of the leukoma with lamellar corneal allograft. Visual acuity was 12/20. Follow-up period was 16 years; c) The Iakymenko-Golubenko KPro, also referred to as the "universal dismountable" KPro; developed and being used at Filatov Inst. since 1978.

Although KPros are well-retained and the success rate is improving, several KPro models still suffer from the drawbacks of complex implantation procedures and serious complications, including retroprosthetic membrane formation, calcification, infection, glaucoma, and retinal detachment. Their use is therefore limited to cases in which allogeneic tissue has failed repeatedly or is contraindicated.

8.2.2 Artificial Corneas as Regeneration Templates

The first artificial cornea that was designed as a 'regeneration template' to promote regeneration of cornea tissue and nerves was reported in Fagerholm et al. (2010) [6]. We successfully completed the first-in-man Phase I clinical study in which cell-free biomimetic corneal implants made from 1-ethyl-3-(3-dimethylaminopropyl)carbodiimide (EDC)/N-hydroxysuccinimide (NHS) crosslinked recombinant human collagen type III (RHCIII) were used to replace the pathologic anterior cornea of ten patients with significant vision loss (from advanced keratoconus and scarring). Once in the patient, the cell-free implants stimulated the patient's own endogenous corneal cells to migrate into the scaffolds, regenerate, producing corneal tissue and nerve regeneration and vision improvement. On its own, the human cornea is unable to regenerate. The implants (marked by arrows) have stably integrated with host tissues (Figure 8.2) for over 4 years and resemble normal,

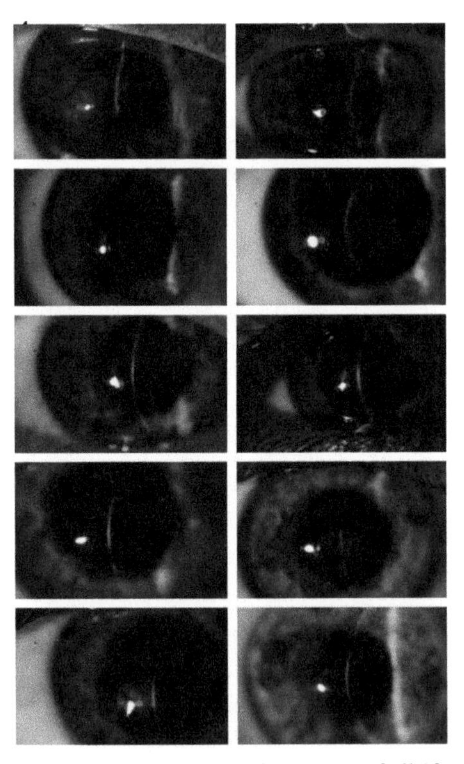

Figure 8.2 Slit lamp biomicroscopy images of the corneas of all 10 patients at 4 years after grafting with a biosynthetic implant. Reproduced from Fagerholm et al. [6].

healthy corneas [7]. There were no signs of rejection and no need for long-term steroid immunosuppression that is required for donor transplantation. In contrast, one control donor allograft corneal implant out of 9 grafts showed rejection within the first year (11%), as per Swedish documented numbers in the cornea graft registry) [1].

However, despite this very promising, positive outcome, several issues still remained. The implants were fairly soft and the mattress sutures used to stabilize the implants sometimes left marks on the implant surface resulting in irregular astigmatism, and sometimes opacities requiring patients to use rigid contact lenses. In addition, the human cornea is avascular and needs to maintain avascularity for transparency and vision. In a rabbit alkali burn model, which recreates a severe cornea inflammation leading to immune privilege loss and neovascularization, the RHCIII implants became neovascularized [8].

It should be noted that the regeneration templates and prostheses are not mutually exclusive. Each has its own usefulness in different transplantation indications.

8.3 Reinforced Collagen Corneal Implants

8.3.1 Interpenetrating Networks of Collagen-Phosphorylcholine as Implants

To address the above issues, we reinforced the EDC/NHS crosslinked porcine collagen or RHCIII implants with a second network of (2-methacryloyloxyethyl phosphorylcholine (MPC) crosslinked with PEGDA. MPC is a synthetic phosphorylcholine lipid that is being used as anti-fouling coatings in arterial stents. Both porcine and RHCIII-MPC implants made from two interpenetrating networks (IPNs) of biopolymers were mechanically stronger, stable when exposed to enzymes [9]. RHCIII-MPC was able to resist neovascularization when tested in rabbit alkali burn models of severe pathology, where the previous generation of collagen only implants and donor allograft corneas became vascularized [8]. More recently, three patients with severe corneal pathologies (resulting from chemical burns or a previous graft rejection) were grafted with tectonic grafts of RHCIII-MPC to treat the symptoms of chronic ulceration [10]. The RHCIII-MPC implants restored the damaged stroma allowing for stable re-epithelialization and relief from pain and discomfort due to the ulceration.

8.3.2 RHCIII-MPC Implants in Herpes Simplex Keratitis

We implanted RHIII-MPC implants into the corneas of mice with Herpes Simplex Keratitis (HSK). Herpes simplex virus serotype 1 (HSV-1) infection of the cornea is the leading cause of infectious blindness in the developed world, and a problem in corneal grafting as there is a high risk of rejection, often due to viral reactivation. Once infected, the individuals may harbour latent virus for the rest of their lives [11].

The mouse allogeneic cornea graft model is essentially a rejection model. It allows for comparisons of time to rejection of implanted biomaterials compared to allografts. When RHCIII-MPC implants were compared to allografts in HSK mouse corneas, there was a trend towards the implants being more resistant to rejection, but the numbers were too small to show statistical significance [12].

8.4 Composite Corneal Implants with Peptide and Gene Therapy Capacity

8.4.1 LL-37

Cathelicidins are innate host defence peptides. In humans, there is only one cathelicidin, the18 kDa human cationic antimicrobial protein (hCAP18), of which LL-37 is a 37 amino acid C-terminal peptide domain with active anti-microbial and anti-viral activity [13]. In the eye, LL-37 is expressed by the cornea epithelium and has been reported to have potent anti-viral activity against Herpes Simplex Virus (HSV)-1 [14].

8.4.2 Implants LL-37 Peptide Release

In Lee et al. [12], we released LL-37 from silica nanoparticles (SiNPs). We showed that LL-37 was able to inhibit viral activity in cultures of human corneal epithelial cells at doses of 10–20 μg/ml, but only when the peptide was applied prophylactically (i.e., prior to viral infection of the cells). Once cells were infected with HSV-1, however, LL-37 could only delay but not prevent viral spreading to other cells. We also showed that LL-37 within SiNP showed more sustained release from collagen-MPC hydrogels (Figure 8.3) than free LL-37 alone. Encapsulation of LL-37 with alginate microparticles was not effective in preventing a burst release of the peptide, even though a high amount of LL-37 could be encapsulated. SiNP encapsulation of LL-37 and integration within a collagen hydrogel to form a composite implant

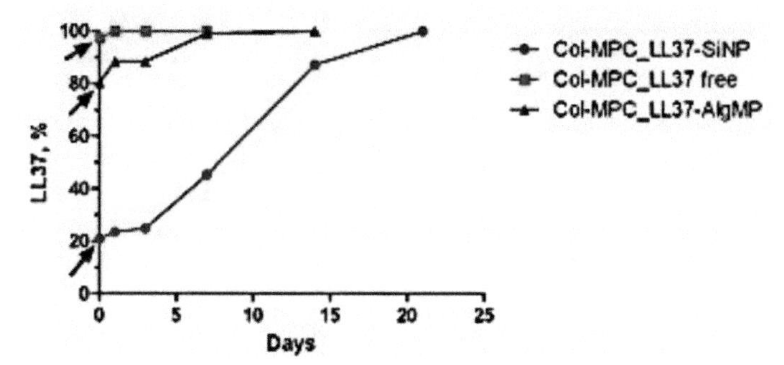

Figure 8.3 Comparison of release profiles of free LL-37, LL-37 within silica nanoparticles (LL-37–SiNP) and alginate microparticles (LL-37–AlgMP) from within collagen-MPC hydrogels.

(Figure 8.4) therefore has the potential for providing sustained release and protection against HSV-1 infection.

It may also be possible to tether LL-37 onto or into hydrogel implants that will be used as grafts in patients with a prior history of HSV-1 infection to prevent reactivation (unpublished data). However, more research and more testing is needed to establish feasibility.

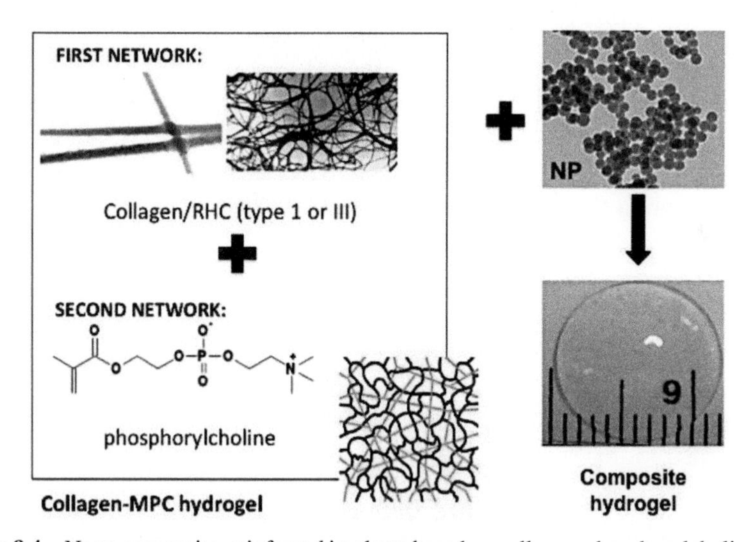

Figure 8.4 Nano-composite, reinforced implants based on collagen-phosphorylcholine interpenetrating networks. Nanoparticles comprising silica dioxide encapsulating LL-37 peptide was evaluated for anti-HSV-1 activity.

8.4.3 Composite Collagen-Cell-Based Implants

We also compared the peptide treatment against gene therapy by transfecting human corneal epithelial cells with the LL-37 gene [12]. The transfected cells expressed and secreted the peptide. The secreted LL-37 inhibited viral binding *in vitro* but in this case, was insufficient to completely protect cells completely from HSV-1 infection. Nevertheless, the secreted LL-37 was able to reduce the incidence of plaque formation and reduced plaque size. The effects were overall weaker that that of exogenously applied LL-37 peptides. It is possible, however, that with further optimization of the gene transfer and copy number of transferred LL-37 genes into the cells, and/or combinations of different antiviral gene sequences, more complete viral resistance can be obtained.

In the future composite grafts releasing LL-37 or some other anti-viral compound or bioactive factor, together with genetically engineered corneal stem cells transfected with the LL-37 gene or another anti-viral gene may together stop HSV-1 activity.

8.5 Conclusion

We have shown that collagen-based hydrogels can be fabricated into corneal implants and used as alternatives to donor corneas for transplantation in some cases. For more severe conditions that have a higher risk of rejection of donor cornea grafts, collagen-hydrogels that are reinforced by additional interpenetrating networks of other biopolymers such as MPC, or by integration of a delivery system of drugs or other bioactives, may in the future become an alternative option to donor corneal grafting.

Ackowledgements

We thank Dr. Stanislav Iakymenko for the photographs used in Figure 8.1.

References

[1] Claesson M, Armitgae WJ, Fagerholm P, Stenevi U. Visual outcome in corneal grafts: A preliminary analysis of the Swedish Corneal Transplant Register. Br J Ophthalmol 2002; 86: 174–80.

[2] Williams KA, Esterman AJ, Bartlett, C. et al. How effective is penetrating corneal transplantation? (Factors influencing long-term outcome in multivariate analysis). Transplantation 2006; 81: 896–901.

[3] Wolfe RA. Long-term renal allograft survival: a cup half-full and half-empty. Am J Transplant 2004; 4:1215–12164.

[4] Dandona L, Naduvilath TJ, Janarthanan M et al. Survival analysis and visual outcome in a large series of corneal transplants in India. Br J Ophthalmol 1997; 81: 726–731.

[5] Iakymenko S. Forty-five years of keratoprosthesis study and application at the Filatov Institute: a retrospective analysis of 1,060 cases. Int. J. Ophthalmol. 2013; 6:375–80.

[6] Fagerholm P, Lagali NS, Merrett K, Jackson WB, Munger R, Liu Y, Polarek JW, Söderqvist M, Griffith M. (2010) A biosynthetic alternative to human donor tissue for inducing corneal regeneration: 24 month follow-up of a Phase I clinical study. Science Transl. Med. 2010; 2: 46–61.

[7] Fagerholm P, Lagali NS, Ong JA, Merrett K, Jackson WB, Polarek JW, Suuronen EJ, Liu Y, Brunette I, Griffith M. (2014) Stable corneal regeneration four years after implantation of a cell-free recombinant human collagen scaffold. Biomaterials 2014; 35: 2420–2427.

[8] Hackett JM, Lagali N, Merrett K, Edelhauser H, Sun Y, Gan L, Griffith M and Fagerholm P. Biosynthetic corneal implants for replacement of pathologic corneal tissue: performance in a controlled rabbit alkali burn model. Invest Ophthalmol Vis Sci 2011; 52: 651–657.

[9] Liu W, Deng C, McLaughlin CR, Fagerholm P, Watsky MA, Heyne B, Scaiano JC, Lagali NS, Munger R, Li F, Griffith M. Collagen-phosphorylcholine interpenetrating network hydrogels as corneal substitutes. Biomaterials 2009; 30: 1551–1559.

[10] Buznyk O, Pasyechnikova N, Islam MM, Iakymenko S, Fagerholm P and Griffith M (2015) Bioengineered Corneas Grafted as Alternatives to Human Donor Corneas in Three High Risk Patients. Clin Transl Sci 8: 558–562.

[11] Choudhary A, Higgins GT, Kaye SB. Herpes Simplex Keratitis and Related Syndromes. In: Reinhard T, Larkin F, eds. Cornea and External Eye Disease: Springer Berlin Heidelberg; 2008:115–152.

[12] Lee CJ, Buznyk O, Kuffova L, Rajendran V, Forrester JV, Phopase J, Islam MM, Skog MM, Ahlqvist J and Griffith, M. (2014) Cathelicidin LL-37 and HSV-1 corneal infection: Peptide versus gene therapy. Transl Vis Sci Technol 2014; 3: 4.

[13] Dürr UHN, Sudheendra US, Ramamoorthy A. LL-37, the only human member of the cathelicidin family of antimicrobial peptides. Biochimica et Biophysica Acta (BBA) – Biomembranes 2006; 1758:1408–1425.
[14] Gordon YJ, Huang LC, Romanowski EG, Yates KA, Proske RJ, McDermott AM. Human cathelicidin (LL-37), a multifunctional peptide, is expressed by ocular surface epithelia and has potent antibacterial and antiviral activity. *Curr Eye Res.* 2005; 30:385–394.

9

Molecular Mechanisms of Smooth Muscle Cell Differentiation from Adipose-Derived Stem Cell

Fang Wang and Jeppe Emmersen

Laboratory for Stem Cell Research, Aalborg University,
Fredrik Bajers Vej 3B, 9220 Aalborg, Denmark

Abstract

Smooth muscle cell can be differentiated from adipose-derived stem cell via various approaches: mechanical force, growth factor or cytokine stimulation as well as cell coculture. The smooth muscle cell derived from stem cell has been employed as a promising cell source to replace the damaged tissue in cardiovascular, gastrointestinal and bladder diseases. The goal of the review is to summarize recent knowledge of methodologies of smooth muscle cell differentiation from adipose-derived stem cell and possible molecular mechanism.

Keywords: Mesenchymal stem cell, Adipose-derived stem cell, Smooth muscle cell differentiation, Molecular mechanism.

Abbreviations

AA	ascorbic acid
ASC	adipose-derived stem cell
A-SMA	α-smooth muscle actin
AT1	angiotensin II receptor type 1
BMP4	bone morphogenetic protein 4
BMPs	bone morphogenetic proteins
CaM	calmodulin

ECM	extracellular matrix
ERK	extracellular signal-regulated kinase
GPCR	G protein-coupled receptor
IL-1β	interleukin-1 beta
LPA	lysophosphatidic acid
MAPK	mitogen-activated protein kinase
MLCK	myosin light chain kinase
MRTF-A	myocardin-related transcription factor-A
MRTF-B	myocardin-related transcription factor-B
PDGF	platelet-derived growth factor
PKC	protein kinase C
PLCL	lactic acid and ε-caprolactone
S1P	sphingosine 1-phosphate
SMC	smooth muscle cell
SM-MHC	smooth muscle myosin heavy chain
SPC	sphingosylphosphorylcholine
SRF	serum response factor
SVF	stromal vascular fraction
TGF-β1	transforming growth factor-β1
VSMC	vascular smooth muscle cell.

9.1 Introduction

Smooth muscle cells (SMCs) constitute the wall of blood vessels, gastrointestinal tracts, respiratory tract, bladder and uterus, thus playing a critical role in homeostasis and a number of physiological processes. Degradation of functional SMC or switching of phenotype is associated with various diseases including atherosclerosis, hypertension, urinary and faecal incontinence [1]. Nowadays, cell-based treatment has been a novel solution involving SMC-related diseases; thus, obtaining fresh functional SMC from stem cells is a critical step in regeneration medicine and tissue engineering involving smooth muscle.

Although embryonic stem cells and born marrow-derived mesenchymal stem cells (BM-MSC) have been extensively investigated, there are still some limitations for their practical clinical application including ethical consideration, low stem cell numbers, painful procedure and donor site morbidity. In contrast, adipose-derived stem cells (ASCs) have been an attractive stem cell source due to its easy accessibility, abundant quantities and no ethical issues. ASC possesses high proliferation capacity allowing rapid expansion *in vitro* and have been differentiated into adipose tissue, bone, cartilage, smooth muscle cell, cardiac muscle, endothelial cells and even neurons [2].

To date, a number of studies have described the differentiation of SMC from ASC via mechanical forces, cytokine and growth factors stimulation, cell–cell coculture. Replacement of damaged SMC with differentiated SMC in cell-based treatment or tissue engineering has been a novel solution, and the efficacy of differentiated SMC has been evaluated in some animal trials [1]. The aim of this review is to summarize recent knowledge in terms of SMC differentiation from ASC and possible molecular mechanism.

9.2 SMC and ASC Characterization

SMC, as a heterogeneous subpopulation, originates from at least eight progenitors including neural crest, secondary heart field, somites, mesoangioblasts, proepicardium, splanchnic mesoderm, mesothelium and various stem cells [3]. SMCs have two different phenotypes: synthetic phenotype and contractile phenotype. Both visceral and vascular SMCs are not terminally differentiated cells in adult organism and are capable of switching between phenotypes due to changes of environmental cues [4]. Contractile phenotypical SMCs are characterized by the expression of specific contractile proteins including α-smooth muscle actin (α-SMA), SM22, calponin, caldesmon, smooth muscle myosin heavy chain (SM-MHC) and smoothelin, which contributes to perform cell-specific contractile function. In contrast, synthetic SMC exhibits the ability to proliferate, migrate and secrete matrix proteins but are also associated with the loss of contractility [5].

Stromal vessel fraction (SVF) containing ASC is produced when adipose tissue is digested with collagenase and centrifuged to remove mature adipocytes. Abundant number, ease of harvest, less invasiveness and immunocompatibility render it a promising stem cell source for repair of damaged tissue or diseased organs. ASCs are characterized by their negative expression of haematopoietic antigens (CD45, CD31) and positive expression of stromal-associated markers (CD29, CD44). Moreover, it is different from BM-MSC due to its positive expression of CD49d and negative expression of CD106. However, due to the lack of single definitive marker, the identification of ASC still needs to consider multifactorial markers including tissue origin, CD marker profile, self-renewal ability and pluripotency [1, 6].

9.3 Differentiation of SMC from ASC and Possible Molecular Mechanism

SMC forms a heterogeneous population with distinct origins, SMC in the body exhibits distinct states of differentiation, and SMC differentiation is controlled by different intracellular mechanism; therefore, SMC differentiation and phenotype switching *in vivo* are extremely complex process [3]. The differentiation is determined by numerous local environmental cues and extrinsic factors including oxygen tension, mechanical influences, cell–cell contact, cell–extracellular matrix (ECM) interactions, humoral factors and neurotransmitters [4, 5]. Among these affecting factors, biochemical factors associated with signalling pathways are undoubtedly a critical component controlling SMC differentiation. Several groups have used a number of different protocols to drive the differentiation of ASC towards a smooth muscle-like cell type, exhibiting both similar morphology and gene and protein expression profiles characteristic of smooth muscle as well as contractility induced by muscarinic agents. An example of changes in morphology is seen in Figure 9.1.

Rodríguez et al. [7] used 100 unit/ml heparin in medium MCDB131 for 6 weeks to successfully drive human ASC to differentiate into phenotypic and functional SMC [7]. This result was confirmed by our group showing when human ASC was cultured in smooth muscle inductive medium (medium MCDB 131 containing 100 unit/ml heparin) for 6 weeks, SMC-specific markers were enhanced after 4 weeks of induction, but decreased from

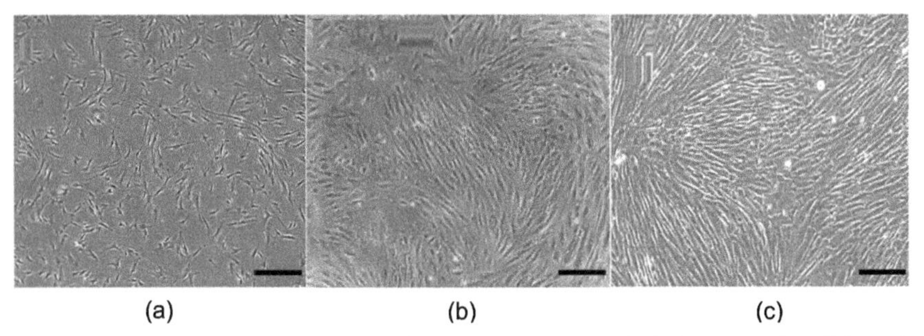

Figure 9.1 Representative images of adipose-derived stem cells before and after differentiation with treatment of TGF-β1 (5 ng/ml) and BMP4 (2.5 ng/ml) in combination for 2 weeks as well as human aortic smooth muscle cells showing the morphological changes. (a) ASC exhibits fibroblast-like shape (b) ASC acquires the spindle-like morphology and the typical "hill and valley" pattern after differentiation 2 weeks similar to (c) human aortic smooth muscle cells. Scale bar, 50 μm for all images.

week 4 to week 6 [8]. Another study indicated that heparin (6–3,200 μg/ml) induced changes in SMC contractile phenotype in dose-dependent fashion [9]. Inhibitory effect of heparin on the SMC proliferation has been extensively investigated: heparin can bind directly to the SMC surface or bind to growth factors of cell surface, thus inhibiting cellular proliferation [10]. Savage et al. [11] employed anti-heparin receptor antibodies to disclose the involvement of heparin receptor on vascular smooth muscle cell growth, which showed anti-heparin receptor antibodies to decrease mitogen-activated protein kinase (MAPK) activity levels after activation in a manner similar to heparin resulting in inhibition of SMC proliferation [11]. In general, vascular SMC proliferation and differentiation are two independent and opposite processes [12]. Then, it is reasonable to speculate that heparin has a great potential to promote ASC to express SMC-like specific contractile proteins by means of binding to growth factors of ASC cell surface, thereby inhibiting ASC growth, although the exact mechanism by which heparin can promote contractile gene expression of SMC has yet to be resolved.

A number of studies have shown the family of transforming growth factor (TGF)-related proteins to be the most potent soluble growth factor-promoting SMC differentiation. Bone morphogenetic proteins (BMPs) represent the largest group in TGF cytokine superfamily [12, 13]. The combination of TGF-β1 (5 ng/ml) and BMP4 (2.5 ng/ml) stimulation for 1 week drove the ASC into mature contractile SMC [14]. Similarly, TGF-β1 and BMP4 were shown to reduce vascular smooth muscle cell (VSMC) proliferation and migration and promote expression of VSMC contractile genes. Additionally, 5 ng/mL TGF-β1 along with 50 ng/mL platelet-derived growth factor (PDGF)-BB or TGF-β1 alone (2 ng/ml for 3 weeks) enhanced SMC markers expression in ASC [15, 16]. It was shown that TGF-β1 proteins modulated SMC differentiation by directly binding to type-I receptor and thereafter activating downstream signals of Smad proteins. Activated Smad2 cooperated with serum response factor (SRF) and myocardin to induce expression of smooth muscle specific genes [14]. Because BMP4 belongs to the TGF-β superfamily, it is reasonable to speculate the activating effects of BMP4 on SMC differentiation might due to induction of TGF-β ligands or activate TGF-β type-I receptor. But the study from Lagna et al. provided evidence that SMC phenotype switch induced by BMP4 from synthetic to contractile was Smad and RhoA/Rho kinase-dependent and TGF-β receptor-independent signaling pathway. The BMP pathway activated transcription of SMC genes by inducing nuclear translocation of the transcription factors myocardin-related transcription factor-A (MRTF-A) and myocardin-related transcription factor-B (MRTF-B), binding the CArG box of specific gene promoters. Therefore, either TGF-β1 or BMP4 has a capability to induce SMC contractile genes. In combination, they may exert a synergistic influence on differentiation through two independent, but crosstalk pathways [17].

Sphingosine 1-phosphate (S1P), sphingosylphosphorylcholine (SPC) and lysophosphatidic acid (LPA) are all natural bioactive lysophospholipids with similar chemical structure containing one long hydrocarbon chain on a three-carbon backbone containing a phosphate group. They can activate specific G protein-coupled receptor (GPCR) superfamily on the membrane [18]. S1P (from 100 nM to 5 μM) and SPC (from 1 to 10 μM) can stimulate differentiation of ASC towards SMC [19]. Another study indicated that ASC cultured in media containing 2 μM SPC for 4 days exhibited a SMC-like contractile phenotype [15]. Cancer-derived LPA induces expression of α-SMA in human ASC after 4 days of stimulation [20].

Since these three phospholipids have similar structure, it is likely to have similar mechanism instructing the ASC differentiation along SMC lineage. With respect to SPC, on one hand, it activated GPCR (Gi/o) extracellular signal-regulated kinase (ERK)-dependent pathway, which stimulated the secretion of TGF-β isoforms inducing late activation of Smad2 through TGF-β type-I receptor kinase. Activated Smad2 cooperated with SRF and myocardin to induce expression of smooth muscle-specific genes [21]. On the other hand, SPC induced RhoA/Rho kinase-dependent nuclear translocation of MRTF-A. Normally, MRTFA/B are inactivated in the cytoplasm due to interaction with G-actin, RhoA/Rho kinase-mediated actin polymerization can make MRTF free from G-actin and enter the nucleus to stimulate SRF-dependent transcription of SMC-specific genes [22]. As to S1P, GPCR receptors S1P2 and S1P3 are relatively important for myogenic differentiation; in addition, S1P has been shown to cross-activate TGF-β signaling pathway in renal mesangial cells and thereby mimicing TGF-β-induced cell responses [23]. Likewise, LPA stimulated differentiation of ASC to myofibroblast-like cells by activating autocrine TGF-β1-Smad signaling pathway as shown by Jeon et al. [20]. These studies suggested that differentiation process of SMC induced by S1P, SPC or LPA might be in part associated with the TGF-β signaling pathway.

Angiotensin II-induced contraction of smooth muscles are involved in multiple pathways including the activation of angiotensin II receptor type 1 (AT1), Ca^{2+} influx, protein kinase C (PKC), MAPK and Rho kinase [24]. Kim et al. also demonstrated that both angiotensin II and bradykinin drove ASC differentiation into contractile SMC phenotype through ERK-dependent activation of the autocrine of TGF-β1-Smad2 crosstalk pathway [25]. Mimetic thromboxane A2 induced differentiation of human ASC to contractile smooth muscle-like cells through calmodulin (CaM)/myosin light chain kinase (MLCK) and RhoA-Rho kinase-dependent actin polymerization [26]. Additionally, it was reported that interleukin-1 beta (IL-1β)-activated macrophages were capable of differentiating ASC into SMC by means of a prostaglandin F2α-mediated paracrine mechanism, involving extracellular signal-regulated kinase, Smad2 and myocardin pathways [27]. All these cytokines belong to vasoactive factors; although upstream signal molecular is varied depending on cytokine varieties, their downstream signal pathways are involved in either of TGF-β1-Smad2 crosstalk pathway or RhoA-Rho kinase-dependent actin polymerization.

Obtaining myocytes from ASC through mechanical stimulation or coculture methods has been tried in some studies but most publications concluded that myogenic markers rather than SMC-specific markers were enhanced via a variety of mechanical stimulation [28]. Other studies employed a combination of methods consisting of cyclic strain and cytokines to increase the SMC-specific marker expression: for instance, when ASC was stimulated with uniaxial cyclic strain (10% strain at 1Hz) with or without 1ng/ml TGF-β1 for 7 days, strain alone decreased the expression of early SMC markers α-SMA and *h1*-calponin in ASC; however, cyclic strain combination with cytokines increased the SMC marker expression [29]. Similarly, ASC seeded on electrospun lactic acid and ε-caprolactone (PLCL) scaffold was subjected to combined strain and varying biochemical effects for 2 weeks (5% uniaxial strain,1HZ; 1 ng/ml TGF-β1; 50 μM β-mercaptoethanol and 0.3 mM ascorbic acid (AA); 1 μM all-*trans* retinoic acid; 10 ng/ml PDGF-BB; 1 μM angiotensin II), and α-SMA and MHC expressions were significantly increased by combined effects with retinoic acid, AA or TGF-β1 [30]. With respect to coculture method, only one publication showed that primary rat bladder smooth muscle cells cocultured with ASC for 2 weeks induced ASC differentiation into SMC, and study indicated that microenvironment cues rather than nuclear fusion induced differentiation [31]. Interesting, distinct pattern of seeding substrate also could change the commitment of stem cell which was verified by one study: uniaxial alignment increased the expression of SM22a and a-SMA, but the other two patterns altering cell direction to different extent promote the expression of chondrogenic and osteogenic markers [28]. These studies suggest that mechanical strain, additional chemical factors, coculture condition and substrate pattern might result in different cellular responses via respective signalling pathways, although it still remains unclear.

9.4 Conclusion and Perspectives

ASC can be driven into functional SMC via various approaches and has been a preferred stem cell source in the repair of damaged tissue and reconstruction of diseased organs. Potential differentiation mechanism is considerably complicated and is an integration of multiple signaling pathways based on varying differentiation approaches, different biochemical factors and extrinsic environmental cues.

Since the implantation of exogenous SMC has been considered as a promising therapy for injured SMC tissue repair and replacement, an optimal differentiation approach is critical important. Oxygen tension influences not only *in vitro* expansion and differentiation but also *in vivo* efficacy after implantation. In addition, oxygen also modulates the paracrine activities of stem cells leading to changes of various secretory factors, which might indirectly affect the differentiation process [32]. Therefore, controlling the differentiation process via modulating oxygen concentration is important in future applications.

Most studies in this field concentrate on *in vitro* differentiation employing biochemical factors using monolayers of culture cells. But an increasing data show how ECM substrate markedly affects SMC phenotypic shift. It has been reported that laminin and collagen type-IV substrates inhibit cell proliferation and increased SMA content and, conversely, substrates of fibronectin and collagen type I enhances SMC proliferation and reduces contractile protein expression [33, 34]. When *in vitro* differentiated SMCs are implanted into the distinct tissues or organs, SMCs have to adapt to a varied 3D physiological structure and a possibly unfavourable environment; therefore, integration of more relative factors to mimic *in vivo* environment is still a major challenge in smooth muscle tissue engineering.

9.5 Conflict of Interest

The authors declare that they have no conflict of interest.

References

[1] De Villiers, J. A., Houreld, N., and Abrahamse, H. (2009). Adipose derived stem cells and smooth muscle cells: implications for regenerative medicine. *Stem Cell Rev. Reports* 5, 256–265.

[2] Strem, B. M. et al. (2005). Multipotential differentiation of adipose tissue-derived stem cells. *Keio J. Med.* 54, 132–141.

[3] Guo, X. et al. (2013). A novel *in vitro* model system for smooth muscle differentiation from human embryonic stem cell-derived mesenchymal cells. *Am. J. Physiol. Cell Physiol.* 304, C289–C298.

[4] Owens, G. K., Kumar, M. S., and Wamhoff, B. R. (2004). Molecular regulation of vascular smooth muscle cell differentiation in development and disease. *Physiol. Rev.* 84, 767–801.

[5] Davis-Dusenbery, B. N., Wu, C., and Hata, A. (2011). Micromanaging vascular smooth muscle cell differentiation and phenotypic modulation. *Arterioscler. Thromb. Vasc. Biol.* 31, 2370–2377.

[6] Zuk, P. (2013). Adipose-derived stem cells in tissue regeneration: a review. *Stem Cells* 2013, 1–35.

[7] Rodríguez, L. V. et al. (2006). Clonogenic multipotent stem cells in human adipose tissue differentiate into functional smooth muscle cells. *Proc. Natl. Acad. Sci. U. S. A.* 103, 12167–12172.

[8] Maeda, Y., Zachar, V., and Emmersen, J. (2010). Variation in cellular expression patterns during differentiation of adipose derived stem cell to smooth muscle cell. *18th United European Gastroenterology Week* 2010, A108.

[9] Beamish, J. A. et al. (2010). The effects of heparin releasing hydrogels on vascular smooth muscle cell hhenotype. *Biomaterials* 30, 6286–6294.

[10] Horiuchi, A. et al. (1999). Heparin inhibits proliferation of myometrial and leiomyomal smooth muscle cells through the induction of alpha-smooth muscle actin, calponin h1 and p27. *Mol. Human Reprod.* 5, 139–145.

[11] Savage, J. M. et al. (2001). Antibodies against a putative heparin receptor slow cell proliferation and decrease MAPK activation in vascular smooth muscle cells. *J. Cell Physiol.* 187, 283–93.

[12] Shi, N., and Chen, S.-Y. Mechanisms simultaneously regulate smooth muscle proliferation and differentiation. *J. Biomed. Res.* 28, 40–46 (2014).

[13] Lagna, G. et al. (2007). Control of phenotypic plasticity of smooth muscle cells by bone morphogenetic protein signaling through the myocardin-related transcription factors. *J. Biol. Chem.* 282, 37244–37255.

[14] Wang, C. et al. (2010). Differentiation of adipose-derived stem cells into contractile smooth muscle cells induced by transforming growth factor-beta1 and bone morphogenetic protein-4. Tissue *Eng. A* 16, 1201–1213.

[15] Harris, L. J. et al. (2011). Differentiation of adult stem cells into smooth muscle for vascular tissue engineering. *J. Surg. Res.* 168, 306–314.
[16] Yang, P. et al. (2008). Experiment of adipose derived stem cells induced into smooth muscle cells. *Chinese J. Reparative Reconst. Surgery* 22, 481–486.
[17] Lagna, G. et al. (2012). Control of phenotypic plasticity of smooth muscle cells by bone morphogenetic protein signaling through the myocardin-related transcription factors. *J. Biol. Chem.* 282, 617–636.
[18] Contos, J. J., Ishii, I., and Chun, J. (2000). Lysophosphatidic acid receptors. *Mol. Pharmacol.* 58, 1188–1196.
[19] Nincheri, P. et al. (2009). Sphingosine 1-phosphate induces differentiation of adipose tissue-derived mesenchymal stem cells towards smooth muscle cells. *Cell. Mol. Life Sci.* 66, 1741–1754.
[20] Jeon, E. S. et al. (2008). Cancer-derived lysophosphatidic acid stimulates differentiation of human mesenchymal stem cells to myofibroblast-like cells. *Stem Cells* 26, 789–797.
[21] Jeon, E. S. et al. (2006). Sphingosylphosphorylcholine induces differentiation of human mesenchymal stem cells into smooth-muscle-like cells through a TGF-β-dependent mechanism. *J. Cell Sci.* 119, 4994–5005.
[22] Jeon, E. S. et al. (2008). A Rho kinase/myocardin-related transcription factor-A–dependent mechanism underlies the sphingosylphosphorylcholine-induced differentiation of mesenchymal stem cells into contractile smoooth muscle cells. *Circulation Res.* 103, 635–642.
[23] Xin, C. Y. et al. (2004). Sphingosine 1-phosphate cross-activates the Smad signaling cascade and mimics transforming growth factor-beta-induced cell responses. *J. Biol. Chem.* 279, 35255–35262.
[24] Rattan, S., Puri, R. N., and Fan, Y.-P. (2003). Involvement of rho and rho-associated kinase in sphincteric smooth muscle contraction by angiotensin II. *Experimental Biol. Med.* 228, 972–981.
[25] Kim, Y. M. et al. (2008). Angiotensin II-induced differentiation of adipose tissue-derived mesenchymal stem cells to smooth muscle-like cells. *Int. J. Biochem. Cell Biol.* 40, 2482–2491.
[26] Kim, M. R. et al. (2009). Thromboxane A2 induces differentiation of human mesenchymal stem cells to smooth muscle-like cells. *Stem Cells* 27, 191–199.
[27] Fetalvero, K. M. et al. (2006). The prostacyclin receptor induces human vascular smooth muscle cell differentiation via the protein kinase A pathway. *Am. J. Physiol. Heart Circulation Physiolol.* 290, H1337–H1346.

[28] Qu, X. et al. (2013). Relative impact of uniaxial alignment vs. form-induced stress on differentiation of human adipose derived stem cells. *Biomaterials* 34, 9812–9818.

[29] Lee, W.-C. C. et al. (2007). Effects of uniaxial cyclic strain on adipose-derived stem cell morphology, proliferation and differentiation. *Biomech. Model. Mechanobiol.* 6, 265–273.

[30] Park, I. S. et al. (2012). Synergistic effect of biochemical factors and strain on the smooth muscle cell differentiation of adipose-derived stem cells on an elastic nanofibrous scaffold. *J. Biomater. Sci.* 23, 1579–1593.

[31] Zhang, R. et al. (2012). Nuclear fusion-independent smooth muscle differentiation of human adipose-derived stem cells induced by a smooth muscle environment. *Stem Cells* 30, 481–490.

[32] Das, R. et al. (2010). The role of hypoxia in bone marrow-derived mesenchymal stem cells: Considerations for regenerative medicine approaches. *Tissue Eng. B* 16, 159–168.

[33] Thyberg, J., and Hultgårdh-Nilsson, A. (1994). Fibronectin and the basement membrane components laminin and collagen type IV influence the phenotypic properties of subcultured rat aortic smooth muscle cells differently. *Cell Tissue Res.* 276, 263–271.

[34] Hedin, U. et al. (1988). Diverse effects of fibronectin and laminin on phenotypic properties of cultured arterial smooth muscle cells. *J. Cell Biol.* 107, 307–319.

[35] Stegemann, J. P., and Nerem, R. M. (2003). Altered response of vascular smooth muscle cells to exogenous biochemical stimulation in two-and three-dimensional culture. *Exp. Cell Res.* 283, 146–155.

Index

About the Editors

Cristian Pablo Pennisi holds a degree in Bioengineering from the National University of Entre Rios (Argentina) and a M.Sc. in Bioelectronics from the CINVESTAV-IPN (Mexico). In 2008 he earned the Ph.D. degree in Biomedical Engineering at Aalborg University (Denmark), place where he continued as postdoctoral fellow at the Laboratory for Stem Cell Research. Since 2011 he is Associate Professor at the Department of Health Science and Technology. His current research work is within the fields of tissue engineering and biomaterials. His main research focus is on studying the effects of mechanical loading and nanoscale topography on cellular fate. Other research interests are in the fields of cellular electrophysiology and biocompatibility of neural interfaces. He is author of several publications in the area of cellular bioengineering. He has served as reviewer of various peer-reviewed journals and conference proceedings in the area of biomedical engineering. Dr. Pennisi is a member of the IEEE Engineering in Medicine and Biology Society and the Danish Stem Cell Society.

Dr. Mayuri Sinha Prasad is a Post-Doctoral fellow at Indiana University, USA. She has done her Ph.D. from Aalborg University, Denmark. Her current research focus encompasses defining Estrogen Receptor alpha (ERα) cistrome that is unique in normal breast epithelial cells and also investigating molecular signaling networks that regulate Estrogen Receptor function in normal and malignant breast epithelial cells.

Her other research interests include cell line based studies for the evaluation of chemosensitivity of human B-cell cancer towards Bendamustine and identification of resistance gene signatures (REGs) as a tool to develop pre-clinical predictive models. She has also worked extensively on the molecular mechanisms underlying proliferation and differentiation of embryonic and adipose derived stem cells cultured in hypoxia.

Dr. Prasad has also co-edited a book titled Innovative Strategies in Tissue Engineering also published by River Publishers in 2015.

Dr. Pranela Rameshwar is a professor of Medicine, Division of Hematology and Oncology at the Rutgers New Jersey Medical School. She received a B.S. degree in medical microbiology from the University of Wisconsin at Madison and a Ph.D. in biology from Rutgers University, New Jersey. Dr. Rameshwar performed postdoctoral studies in hematopoiesis at New Jersey Medical School. Thereafter, she became a faculty member in the same department.

Dr. Rameshwar's research interest is in the translation of stem cell, including drug delivery with stem cells. Her laboratory also studies breast cancer dormancy with a focus on cancer stem cells; neural regulation of hematopoiesis and the immunology of adult human mesenchymal stem cells. Her research continues to be funded by federal, state, and other agencies. Dr. Rameshwar has trained several doctoral students, including those in the physician scientist track. In addition, Dr. Rameshwar mentors junior faculty members.

Dr. Rameshwar has authored >200 publications, which include original articles, reviews, editorials and book chapters. She has also edited books. Dr. Rameshwar is also involved in teaching. She initiated a certificate program in Stem Cell Biology and directs four graduate level courses in stem cell biology. Dr. Rameshwar is also a university Master Educator. Dr. Rameshwar also serves on several grant review panels both nationally and internationally. She has given numerous invited seminars both nationally and internationally.